Lecture Notes in Statistics 131

Edited by P. Bickel, P. Diggle, S. Fienberg, K. Krickeberg, I. Olkin, N. Wermuth, S. Zeger

Springer
New York
Berlin
Heidelberg
Barcelona
Budapest
Hong Kong
London
Milan
Paris
Santa Clara
Singapore
Tokyo

Joel L. Horowitz

Semiparametric Methods in Econometrics

Joel L. Horowitz
Department of Economics
University of Iowa
Iowa City, IA 52242

Library of Congress Cataloging-in-Publication Data
Horowitz, Joel.
Semiparametric methods in econometrics / Joel L. Horowitz.
p. cm. -- (Lecture notes in statistics ; 131)
Includes bibliographical references and index.
ISBN 0-387-98477-1 (softcover : alk. paper)
1. Econometrics. 2. Estimation theory. I. Title. II. Series:
Lecture notes in statistics (Springer-Verlag) ; v. 131.
HB139.H65 1998
330'.01'5195--dc21 98-11300
Printed on acid-free paper.

Camera ready copy provided by the author.
Printed and bound by Braun-Brumfield, Ann Arbor, MI.
Printed in the United States of America.

9 8 7 6 5 4 3 2 1

ISBN 0-387-98477-1 Springer-Verlag New York Berlin Heidelberg SPIN 10659542

To N, S, and K

Preface

This book is based on a series of lectures on semiparametric estimation that I gave at the Paris-Berlin Seminar in Garchy, France, in October 1996. The topics that are covered in the book are the same as those of the lectures, but the presentation in the book is much more detailed. Some of the material is new and is presented here for the first time.

Several people deserve special thanks for helping to make this book possible. I thank Wolfgang Härdle for arranging the invitation to the Paris-Berlin Seminar that led to the book. Wolfgang Härdle, Ruud Konning, George Neumann, and Harry Paarsch read drafts of the manuscript. I am deeply indebted to them for comments and suggestions that greatly improved the presentation. I also thank John Kimmel for all his help and advice on the details of production. Partial financial support was provided by the National Science Foundation under grant SBR-9617925. Finally, I thank my wife, Susan, for her support and patience during both the writing of the book and the research that made it possible.

Contents

Chapter 1
Introduction

Many estimation problems in econometrics involve an unknown function as well as an unknown finite-dimensional parameter. Models and estimation problems that involve an unknown function and an unknown finite-dimensional parameter are called *semiparametric*.

There are many simple and familiar examples of semiparametric estimation problems. One is estimating the vector of coefficients β in the linear model

$$Y = X\beta + U,$$

where Y is an observed dependent variable, X is an observed (row) vector of explanatory variables, and U is an unobserved random variable whose mean conditional on X is zero. If the distribution of U is known up to finitely many parameters, then the method of maximum likelihood provides an asymptotically efficient estimator of β and the parameters of the distribution of U. Examples of finite-dimensional families of distributions are the normal, the exponential, and the Poisson. Each of these distributions is completely determined by the values of one or two constants (e.g., the mean and the standard deviation in the case of the normal distribution). If the distribution of U is not known up to finitely many parameters, the problem of estimating β is semiparametric. The most familiar semiparametric estimator is ordinary least squares, which is consistent under mild assumptions regardless of the distribution of U. By contrast, a parametric estimator of β need not be consistent. For example, the maximum likelihood estimator is inconsistent if X is exponentially distributed but the analyst erroneously assumes it to be lognormal.

The problem of estimating the coefficient vector in a linear model is so simple and familiar that labeling it with a term as fancy as *semiparametric* may seem excessive. A more difficult problem that has received much attention in econometrics is semiparametric estimation of a binary-response model. Let Y be a random variable whose only possible values are 0 and 1, and let X be a vector of covariates of Y. Consider the problem of estimating the probability that $Y = 1$ conditional on X. Suppose that the true conditional probability is

$$P(Y = 1|X = x) = F(x\beta),$$

where F is a distribution function and β is a vector of constant parameters that is conformable with X. If F is assumed to be known *a priori*, as in a binary probit model, where F is the standard normal distribution function, the only problem is to estimate β. This can be done by maximum likelihood. F is rarely known in applications, however. If F is misspecified, then the maximum likelihood estimators of β and $P(Y=1|X=x)$ are inconsistent except in special cases, and inferences based on them can be highly misleading. In contrast to estimation of a mean, where the simplest and most familiar estimator is automatically semiparametric, the distribution function F has a non-trivial influence on the most familiar estimator of a binary-response model.

Many other important estimation problems involve unknown functions in non-trivial ways. Often, as is the case in the foregoing examples, the unknown function is the distribution function of an unobserved random variable that influences the relation between observed variables. As will be discussed later in this book, however, the unknown function may also describe other features of a model. The methods needed to estimate models that include both an unknown function and an unknown finite-dimensional parameter (semiparametric models) are different from those needed to estimate models that contain one or more unknown functions but no finite-dimensional parameters. Thus, it is important to distinguish between the two types of models. Models or the latter type are called *nonparametric*. This book is concerned with estimation of semiparametric models.

Semiparametric estimation problems have generated large literatures in both econometrics and statistics. Most of this literature is highly technical. Moreover, much of it is divorced from applications, so even technically sophisticated readers can have difficulty judging whether a particular technique is likely to be useful in applied research.

This book aims at mitigating these problems. I have tried to present the main ideas underlying a variety of semiparametric methods in a way that will be accessible to graduate students and applied researchers who are familiar with econometric theory at the level found (for example) in the textbooks by Amemiya (1985) and Davidson and MacKinnon (1993). To this end, I have emphasized ideas rather than technical details and have provided as intuitive an exposition as possible. I have given heuristic explanations of how important results are proved, rather than formal proofs. Many results are stated without any kind of proof, heuristic or otherwise. In all cases, however, I have given references to sources that provide complete, formal proofs.

I have also tried to establish links to applications and to illustrate the ability of semiparametric methods to provide insights about data that are not readily available using more familiar parametric methods. To this end, each chapter contains a real-data application as well as examples without data of applied problems in which semiparametric methods can be useful.

I have not attempted to provide a comprehensive treatment of semiparametric methods in econometrics. The subject is so large that any effort to treat it comprehensively would require either a book many hundreds of pages longer than this one or a highly abstract presentation. Accordingly, this book treats only a small set of estimation problems that I have selected because they are important in applied econometrics and are closely related to my own research. Some other important estimation problems are described briefly, but I have not attempted to list all important problems. Semiparametric estimation of models for time series is not treated. The subject of specification testing, which has received much recent attention, is also not treated. Treatments of semiparametric estimation that are more comprehensive though also more abstract are available in the book of Bickel *et al.* (1993) and the review by Powell (1994).

Chapter 2
Single-Index Models

One of the most important tasks of applied econometrics and statistics is estimating a conditional mean function. For example, one may want to estimate the mean annual earnings of workers in a certain population as a function of observable characteristics such as level of education and experience in the workforce. As another example, one may want to estimate the probability that an individual is employed conditional on observable characteristics such as age, level of education, and sex.

The methods available for estimating a conditional mean function and the results that these methods produce depend critically on what one assumes is known *a priori* about the population or process that generates the data. This chapter describes a method called *single-index modeling*. Single-index models relax some of the restrictive assumptions of familiar parametric models of conditional mean functions, such as linear models and binary probit models. At the same time, single-index models maintain many of the desirable features of linear models and least-squares methods. In particular, single-index models are often easy to compute, and their results are easy to interpret.

2.1 Definition of a Single-Index Model

Let Y be a scalar random variable and X be a $1\times k$ random vector. Let E denote the expectation operator. Let x be a specific value of X. The conditional mean function, $E(Y|x)$, gives the mean of Y conditional on $X = x$ for all possible values of x. In a semiparametric single-index model, the conditional mean function has the form

$$E(Y|x) = G(x\beta), \tag{2.1}$$

where β is an unknown $k\times 1$ constant vector and G is an unknown function. The quantity $x\beta$ is called an *index*. The inferential problem in (2.1) is to estimate both from observations of (Y, X).

Model (2.1) contains many widely used parametric models as special cases. If G is the identity function, then (2.1) is a linear model. If G is the cumulative normal or logistic distribution function, then (2.1) is a binary probit or logit model. A Tobit model is obtained if one assumes that G is the conditional mean function of Y in the model

$$Y = \max(0, X\beta + U),$$

where U is an unobserved, normally distributed random variable. When G is unknown, (2.1) provides a specification that is more flexible than a parametric model but retains many of the desirable features of parametric models. This characteristic of semiparametric single-index models is explained in more detail in the next section.

2.2 Why Single-Index Models Are Useful

Because a single-index model does not assume that G is known, it is more flexible and less restrictive than are parametric models for conditional mean functions, such as linear models and binary probit models. Flexibility is important in applications because there is usually little justification for assuming that G is known *a priori*, and seriously misleading results can be obtained if one makes an incorrect specification of G. Use of a semiparametric single-index model reduces the risk of obtaining misleading results. Section 2.9 gives an empirical example of this.

Of course, one may argue that there is little justification for assuming that the conditional mean function has the single-index structure (2.1). In fact, it is possible to estimate the conditional mean function completely nonparametrically. In nonparametric estimation, $E(Y|x)$, considered as a function of x, is assumed to satisfy smoothness conditions (e.g., differentiability), but no assumptions are made about its shape or the form of its dependence on x. Nonparametric estimation techniques and their properties are summarized in the Appendix. Härdle (1990) provides a more complete discussion. Nonparametric estimation of a conditional mean function maximizes flexibility and minimizes (but does not eliminate) the risk of specification error. The price of this flexibility can be high, however, for several reasons.

First, estimation precision decreases rapidly as the dimension of X increases (the so-called curse of dimensionality). Specifically, the fastest achievable rate of convergence in probability of an estimator of $E(Y|x)$ decreases as the number of continuously distributed components of X increases (Stone 1980). As a result, impracticably large samples may be needed to obtain acceptable estimation precision if X is multidimensional, as it often is in economic applications. This is an unavoidable problem in nonparametric estimation.

A single-index model avoids the curse of dimensionality because, as will be seen later in this chapter, the *index* $X\beta$ aggregates the dimensions of x.

Consequently, G in a single-index model can be estimated with the same rate of convergence in probability that it would have if the one-dimensional quantity $X\beta$ were observable. Moreover, β can be estimated with the same rate of convergence, $n^{-1/2}$, that is achieved in a parametric model. Thus, in terms of rate of convergence in probability, the single-index model is as accurate as a parametric model for estimating β and as accurate as a one-dimensional nonparametric mean-regression for estimating G. This dimension-reduction feature of single-index models gives them a considerable advantage over nonparametric methods in applications where X is multidimensional and the single-index structure is plausible.

A second problem with nonparametric estimation is that its results can be difficult to display, communicate, and interpret when X is multidimensional. Nonparametric estimates usually do not have simple analytic forms. If x is one- or two-dimensional, the estimate of $E(Y|x)$ can be displayed graphically. When X has three or more components, however, only reduced-dimension projections of $E(Y|x)$ can be displayed. Many such displays and much skill in interpreting them may be needed to fully convey and comprehend the shape of $E(Y|x)$.

A further problem with nonparametric estimation is that it does not permit extrapolation. That is, it does not provide predictions of $E(Y|x)$ at points x that are not in the support of X. This is a serious drawback in policy analysis and forecasting, whose main purpose often is to make statements about what might happen under conditions that do not exist in the available data. A single-index model, by contrast, permits extrapolation within limits. Specifically, a single-index model yields predictions of $E(Y|x)$ at values of x that are not in the support of X but are in the support of $X\beta$. A parametric model (that is, a model in which $E(Y|x)$ is known up to a finite-dimensional parameter) provides predictions at all values of x.

The differences among the abilities of nonparametric, single-index, and parametric models to support extrapolation reflect the different strengths of the *a priori* assumptions about the data-generation process in each of these models. Extrapolation is possible only if one is willing to make assumptions about the form of the conditional mean function at points beyond the support of the data. These assumptions are, of course, untestable. A nonparametric model makes no such assumptions, and, therefore, provides no capability for extrapolation. A parametric model makes very strong assumptions and, therefore, provides unlimited opportunities for extrapolation if its assumptions are satisfied by the data generation process. A semiparametric model makes assumptions of intermediate strength and provides an intermediate capability for extrapolation.

The characterization of single-index models as *intermediate* is useful more generally. By making assumptions that are weaker than those of a fully parametric model but stronger than those of a nonparametric model, a single-index model reduces the risk of misspecification relative to a parametric model while avoiding some serious drawbacks of fully nonparametric methods such as the curse of dimensionality, difficulty of interpretation, and lack of extrapolation capability.

There is an important exception to the characterization of a single-index model as intermediate or as making weaker assumptions than a nonparametric model. This exception occurs in the estimation of structural economic models. A structural model is one whose components have a clearly defined relation to economic theory. It turns out that the restrictions needed to make possible a structural interpretation of a nonparametric model can cause the nonparametric model to be no more general than a single-index model. To see why, consider a simple structural model of whether an individual is employed or unemployed.

Example 2.1: A Binary-Response Model of Employment Status

An important model in economic theory states that an individual is employed if his market wage exceeds his reservation wage, which is the value of his time if unemployed. Let Y^* denote the difference between an individual's market and reservation wages. Consider the problem of inferring the probability distribution of Y^* conditional on a vector of covariates, X, that characterizes the individual and, possibly, the state of the economy. Let H denote the conditional mean function. That is, $E(Y^*|x) = H(x)$. Then

$$(2.2) \qquad Y^* = H(X) - U ,$$

where U is an unobserved random variable that captures the effects of variables other than X that influence employment status (unobserved covariates). Suppose that U is independent of X, and let F be the cumulative distribution function (CDF) of U. The estimation problem is to infer H and F. It turns out, however, that this problem has no solution unless suitable *a priori* restrictions are placed on H and F. The remainder of this example explains why this is so and compares alternative sets of restrictions.

To begin, suppose that Y^* were observable. Then H could be estimated nonparametrically as the nonparametric mean-regression of Y^* on X. More importantly, the population distribution of the random vector (Y^*,X) would identify (that is, uniquely determine) H if H is a continuous function of the continuous components of X. F would also be identified if Y^* were observable, because F would be the CDF of the identified random variable $U = H(X) - Y^*$. F could be estimated as the empirical distribution function of the quantity that is obtained from U by replacing H with its estimator. However, Y^* is not observable because the market wage is observable only for employed individuals, and the reservation wage is never observable. An individual's employment status is observable, though. Moreover, according to the economic theory model, $Y^* \geq 0$ for employed individuals, whereas $Y^* < 0$ for individuals who are not employed. Thus, employment status provides an observation of the sign of Y^*. Let Y be the indicator of employment status: $Y = 1$ if an individual is employed and $Y = 0$ otherwise. We now investigate whether H and F can be inferred from observations on (Y, X).

To solve this problem, let $G(x) = P(Y = 1|x)$ be the probability that $Y = 1$ conditional on $X = x$. Because Y is binary, $G(x) = E(Y|x)$, so G can be estimated as the nonparametric mean regression of Y on X. More importantly, the population distribution of the observable random vector (Y, X) identifies G if G is a continuous function of the continuous components of X. It follows from (2.2) that $P(Y^* \geq 0|x) = F[H(x)]$. Therefore, since $Y^* \geq 0$ if and only if $Y = 1$, $P(Y^* \geq 0|x) = P(Y = 1|x)$, and

$$(2.3) \qquad F[H(x)] = G(x) .$$

The problem of inferring H and F can now been seen clearly. The population distribution of (Y, X) identifies G. H and F are related to G by (2.3). Therefore, H and F are identified and nonparametrically estimable only if (2.3) has a unique solution for H and F in terms of G.

One way to achieve identification is by assuming that H has the single-index structure

$$(2.4) \qquad H(x) = x\beta .$$

If (2.4) holds, then identification of H is equivalent to identification of β. As will be discussed in Section 2.4, β is identified if X has at least one continuously distributed component whose β coefficient is non-zero, F is differentiable and non-constant, and certain other conditions are satisfied. F is also identified and can be estimated as the nonparametric mean-regression of Y on the estimate of $X\beta$.

The single-index model (2.4) is more restrictive than a fully nonparametric model, so it is important to ask whether H and F are identified and estimable nonparametrically. This question has been investigated by Matzkin (1992, 1994). The answer turns out to be *no* unless H is restricted to a suitably small class of functions. To see why, suppose that X is a scalar and

$$G(x) = \frac{1}{1+e^{-x}}$$

Then one solution to (2.3) is

$$H(x) = x$$

and

$$F(u) = \frac{1}{1+e^{-u}}; \quad -\infty \le u \le \infty .$$

Another solution is

$$H(x) = \frac{1}{1+e^{-x}} .$$

and

$$F(u) = u; \quad 0 \le u \le 1 .$$

Therefore, (2.3) does not have a unique solution, and F and H are not identified unless they are restricted to classes that are smaller than the class of all distribution functions (for F) and the class of all functions (for H).

Matzkin (1992, 1994) gives examples of suitable classes. Each contains some single-index models but none contains all. Thus, the single-index specification consisting of (2.3) and (2.4) contains models that are not within Matzkin's classes of identifiable, nonparametric, structural models. Similarly, there are identifiable, nonparametric, structural models that are not single index models. Therefore, Matzkin's classes of identifiable, nonparametric, structural models are neither more nor less general than the class of single-index models. It is an open question whether there are interesting and useful classes of identifiable, nonparametric, structural models of the form (2.3) that contain all identifiable single-index submodels of (2.3). ■

2.3 Other Approaches to Dimension Reduction

One of the main benefits of a single-index model is its ability to aggregate the dimensions of X and, thereby, to avoid the curse of dimensionality of nonparametric estimation. This section briefly describes several other approaches to dimension reduction and explains their relation to single-index models.

2.3.1 Multiple-Index Models

A multiple-index model has the form

(2.5) $$E(Y|x) = x_0\beta_0 + G(x_1\beta_1, \ldots, x_M\beta_M) ,$$

where $M \geq 1$ is a known integer, x_m ($m = 0,\ldots, M$) is a subvector of x, β_m ($m = 0,\ldots, M$) is a vector of unknown parameters, and G is an unknown function. This model has been investigated in detail by Ichimura and Lee (1991). A slightly different form of the model called *sliced inverse regression* has been proposed by Li (1991). To identify the β vectors in (2.5), each x_i ($i = 1,\ldots, M$) must have at least one continuously distributed component that is not a component of any other vector x_j ($j = 0,\ldots, M$; $j \neq i$) and whose coefficient is non-zero. Because of this requirement, the class of multiple-index models (2.5) and the class of single-index models (2.1) are non-nested. Each class contains models that are not contained in the other, so neither class is more general than the other.

If the β parameters in (2.5) are identified and certain other conditions are satisfied, then the β's can be estimated with a $n^{-1/2}$ rate of convergence in probability, the same as the rate with a parametric model. The estimator of $E(Y|x)$, however, converges at the rate of a nonparametric estimate of a conditional-mean function with an M-dimensional argument. Thus, in a multiple-index model, estimation of $E(Y|x)$ but not of β suffers from the curse of dimensionality as M increases.

The applications in which a multiple-index model is likely to be useful are different from those in which a single-index model is likely to be useful. The curse of dimensionality associated with increasing M and the need to specify identifiable indices *a priori* limits the usefulness of multiple-index models for estimating $E(Y|x)$. There are, however, applications in which the object of interest is β, not $E(Y|x)$, and the specification of indices arises naturally. The following example provides an illustration.

Example 2.2: A Wage Equation with Selectivity Bias

Let W denote the logarithm of an individual's market wage. Suppose we want to estimate $E(W|Z = z) \equiv E(W|z)$, where Z is a vector of covariates such as experience and level of education. Suppose, also, that the conditional mean function is assumed to be linear. Then $E(W|z) = z\alpha$, where α is a vector of coefficients. Moreover,

$$W = Z\alpha + V, \tag{2.6}$$

where V is an unobserved random variable that represents the effects on wages of variables not included in Z (e.g., unobserved ability). If (W,Z) were observable for a random sample of individuals, then α could be estimated, among other ways, by applying ordinary least squares to (2.6). However, W is observable only for employed individuals, and a random sample of individuals is likely to include some who are unemployed. Therefore, unless attention is restricted to groups in which nearly everyone is employed, one cannot expect to observe (W,X) for a random sample of individuals.

To see how this problem affects estimation of α and how it can lead to a multiple-index model, suppose that employment status is given by the single-index model consisting of (2.2) and (2.4). Then the mean of W conditional on $Z = z$, $X = x$, and $Y = 1$ is

$$E(W|z, x, Y = 1) = z\alpha + E(V|z, x, U \leq x\beta). \tag{2.7}$$

If V is independent of Z and X conditional on U, then (2.7) becomes

$$E(W|z, x, Y = 1) = z\alpha + G(x\beta), \tag{2.8}$$

where $G(x\beta) \equiv E(V|z, x, U \leq x\beta)$. Equation (2.8) is a multiple-index model that gives the mean of log wages of employed individuals conditional on covariates Z and X. Observe that (2.8) is not equivalent to the linear model (2.6) unless $E(V|z, x, U \leq x\beta)=0$. If $E(V|z, x, U \leq x\beta) \neq 0$, estimation of (2.6) will give rise to a *selectivity bias* arising from the fact that one does not observe W for a random sample of individuals. This is also called a *sample selection* problem because the observed values of W are selected non-randomly from the population. Gronau (1974) and Heckman (1974) used models like (2.7) under the additional assumption that V and U are bivariate normally distributed. In this case G, is known up to a scalar parameter, and the model is no longer semiparametric.

In (2.8), α is identified only if X has at least one continuously distributed component that is not a component of Z and whose β coefficient is nonzero. The credibility of such an *exclusion restriction* in an application can be highly problematic. Manski (1994, 1995) provides a detailed discussion of the problems of identification in the presence of sample selection. ■

2.3.2 Partially Linear Models

In a partially linear model, the vector of covariates X is divided into two non-overlapping subvectors, X_1 and X_2. The model has the form

$$E(Y|x_1, x_2) = x_1\beta + G(x_2). \tag{2.9}$$

This model, has been investigated in detail by Robinson (1988). Ai and McFadden (1997) present a generalized version. Identification of β requires the strong exclusion restriction that none of the components of X_1 can be exact linear combinations of components of X_2. Among other things, this implies that X_1 and X_2 cannot have any components in common. This exclusion restriction makes the partially linear specification (2.9) more restrictive than the multiple-index specification (2.5). Any partially linear model is trivially a multiple-index model in which each index $x_i\beta_i$ ($i = 1,\ldots, M$) in (2.5) is a single component of x_2 in (2.9). However, not every multiple-index model is partially linear in the sense of (2.9).

The class of partially linear models is distinct from the class of single-index models. A single-index model is not partially linear unless G is a linear function. Furthermore, a partially linear model is a single-index model only in this case.

If β is identified in (2.9) and certain other conditions are satisfied, then β can be estimated with a $n^{-1/2}$ rate of convergence in probability. The estimator of $E(Y|x)$ converges at the rate of a nonparametric estimate of a conditional-mean function whose argument has dimension equal to the number of continuous components of X_2. Therefore, a partially linear model achieves dimension reduction with respect to X_1 but suffers from the curse of dimensionality with respect to X_2.

2.3.3 Additive Models

An additive model of $E(Y|x)$ has the form

(2.10) $$E(Y|x) = g_1(x_1)+\ldots+g_k(x_k),$$

where $k = \dim(X)$, x_i is the i'th component of x, and the g_i are unknown functions. A generalized additive model has the form

(2.11) $$H[E(Y|x)] = g_1(x_1) +\ldots+ g_k(x_k),$$

where H is a known function and the remaining symbols are as in (2.10). It is clear that the class of additive models (2.10) is completely contained in the class of generalized additive models (2.11).

Additive and generalized additive models have been investigated by Andrews and Whang (1990), Breiman and Friedman (1985), Hastie and Tibshirani (1990), Linton (1996), Linton and Härdle (1996), Linton and Nielsen (1995), Newey (1994), and Stone (1985), among others. These models are important because the g_i functions can be estimated nonparametrically with the one-dimensional nonparametric rate of convergence. Thus, (2.10) and (2.11) enable the curse of dimensionality to be avoided without imposing parametric restrictions on the additive components.

The class of generalized additive models is distinct from the class of single-index models. That is, there are generalized additive models that are not single-index models, and there are single-index models that are not generalized additive models for any given choice of H. Thus, neither class is more general than the other.

Generalized additive models are plausible alternatives to single-index models in applications. There are few *a priori* grounds for preferring one type of model to the other. Methods for choosing empirically between the two types of models in applications have not been developed.

2.3.4 Projection-Pursuit Regression

In projection-pursuit regression, $E(Y|x)$ is assumed to have the form

$$(2.12) \qquad E(Y|x) = g_1(x\beta_1) + \ldots + g_M(x\beta_M),$$

where the functions g_m, parameter vectors β_m ($m = 1, \ldots, M$), and M are unknown. This model has been proposed for use in descriptive data analysis (Friedman and Stuetzle 1981, Huber 1985).

Model (2.12) is not identified if the g's, β's, and M are unrestricted. Under suitable identifying restrictions, (2.12) becomes a semiparametric single-index, multiple-index, or additive model. For example, (2.12) is a single-index model if $M = 1$. If $M < \dim(X)$, is specified *a priori*, g_1 is a linear function, and each β_m has a non-zero component whose value is 0 in $\beta_{m'}$ for all $m' \neq m$, then (2.12) is a special case of a multiple-index model. It is an open question whether there are identifying restrictions that yield useful forms of (2.12) that are not single-index, multiple-index, or additive models.

2.4 Identification of Single-Index Models

The remainder of this chapter is concerned with the semiparametric single-index model (2.1).

2.4.1 Conditions for Identification of β and G

Before estimation of β and G can be considered, restrictions must be imposed that insure their identification. That is, β and G must be uniquely determined by the population distribution of (Y, X). Identification of single-index models has been investigated by Ichimura (1993) and, for the special case of binary-response models, by Manski (1988). Some of the restrictions required for identification are easy to see. It is clear that β is not identified if G is a constant function. It is also clear that as in a linear model, β is not identified if there is an exact linear relation among the components of X (perfect multicollinearity). In other words, β is not identified if there are a constant vector α and a constant scalar c such that $X\alpha = c$ with probability one.

To obtain additional conditions for identification, let γ be any constant and δ be any non-zero constant. Define the function G^* by the relation $G^*(\gamma + \delta v) = G(v)$ for all v in the support of $X\beta$. Then

(2.13) $E(Y|x) = G(x\beta)$

and

(2.14) $E(Y|x) = G^*(\gamma + x\beta\delta)$.

Models (2.13) and (2.14) are observationally equivalent. They could not be distinguished empirically even if the population distribution of (Y, X) were known. Therefore, β and G are not identified unless restrictions are imposed that uniquely specify γ and δ. The restriction on γ is called a *location normalization*, and the restriction on δ is called a *scale normalization*. Location normalization can be achieved by requiring X to contain no constant (intercept) component. Scale normalization can be achieved by setting the β coefficient of one component of X equal to one. In this book it will be assumed that the components of X have been arranged so that scale normalization is carried out on the coefficient of the first. Moreover, for reasons that will now be explained, it will also be assumed that this component of X is a continuously distributed random variable.

To see why there must be at least one continuously distributed component of X, consider the following example.

Example 2.3: A Single-Index Model with Only Discrete Covariates

Suppose that $X = (X_1, X_2)$ is two-dimensional and discrete with support consisting of the corners of the unit square: (0,0), (1,0), (0,1), and (1,1). Set the coefficient X_1 equal to one to achieve scale normalization. Then (2.1) becomes

$$E(Y|x) = G(x_1 + \beta_2 x_2).$$

Suppose that the values of $E(Y|x)$ at the points of support of X are as shown in Table 2.1. Then all choices of β_2 and G that equate the entry in the second column to the corresponding entry in the third column are correct models of $E(Y|x)$. These models are observationally equivalent and would be indistinguishable from one another even if the population distribution of (Y, X) were known. There are infinitely many such models, so β_2 and G are not identified. Bierens and Hartog (1988) provide a detailed discussion of alternative, observationally equivalent forms of β and G when all components of X are discrete. ■

Table 2.1: An Unidentified Single-Index Model

(x_1, x_2)	$E(Y\|x)$	$G(x_1 + \beta_2 x_2)$
(0, 0)	0	$G(0)$
(1, 0)	0.1	$G(1)$
(0, 1)	0.3	$G(\beta_2)$
(1, 1)	0.4	$G(1 + \beta_2)$

Another requirement for identification is that G must be differentiable. To understand why, observe that the distinguishing characteristic of a single-index model that makes identification possible is that $E(Y|x)$ is constant if x changes in such a way that $x\beta$ stays constant. However, if $X\beta$ is a continuously distributed random variable, as it is if X has at least one continuous component with a non-zero coefficient, the set of X values on which $X\beta = c$ has probability zero for any c. Events of probability zero happen too infrequently to permit identification. If G is differentiable, then $G(X\beta)$ is close to $G(c)$ whenever $X\beta$ is close to c. The set of X values on which $X\beta$ is within any specified non-zero distance of c has non-zero probability for any c in the interior of the support of $X\beta$. This permits identification of β through "approximate" constancy of $X\beta$.

It is now possible to state a complete set of conditions for identification of β in a single-index model. This theorem assumes that the components of X are all continuous random variables. Identification when some components of X are discrete is more complicated. This case is discussed after the statement of the theorem.

Theorem 2.1 (Identification in Single-Index Models): Suppose that $E(Y|x)$ satisfies model (2.1) and X is a k-dimensional random variable. Then β and G are identified if the following conditions hold:

(a) G is differentiable and not constant on the support of $X\beta$.

(b) The components of X are continuously distributed random variables that have a joint probability density function.

(c) The support of X is not contained in any proper linear subspace of $\Re^k$.

(d) $\beta_1 = 1$. ■

See Ichimura (1993) and Manski (1988) for proofs of several versions of this theorem. It is also possible to prove a version that permits some components of X to be discrete. Two additional conditions are needed. These are: (1) varying the values of the discrete components must not divide the support of $X\beta$ into disjoint subsets, and (2) G must satisfy a non-periodicity condition.

The following example illustrates the need for condition (1).

Example 2.4: Identification of a Single-Index Model with Continuous and Discrete Covariates

Suppose that X has one continuous component, X_1, whose support is $[0,1]$, and one discrete component, X_2, whose support is the two-point set $\{0,1\}$.

Assume that X_1 and X_2 are independent and that G is strictly increasing on [0,1]. Set $\beta_1 = 1$ to achieve scale normalization. Then $X\beta = X_1 + \beta_2 X_2$. Observe that $E(Y|x_1,0) = G(x_1)$ and $E(Y|x_1,1) = G(x_1 + \beta_2)$. Observations of X for which $X_2 = 0$ identify G on [0,1]. However, if $\beta_2 > 1$, the support of $X_1 + \beta_2$ is disjoint from [0,1], and β_2 is, in effect, an intercept term in the model for $E(Y|x_1,1)$. As was explained in the discussion of location and scale normalization, an intercept term is not identified, so β_2 is not identified in this model.

The situation is different if $\beta_2 < 1$, because the supports of X_1 and $X_1 + \beta_2$ then overlap. The interval of overlap is $[\beta_2, 1]$. Because of this overlap, there is a subset of the support of X on which $X_2 = 1$ and $G(X_1 + \beta_2) = G(v)$ for some $v \in [0,1]$. The subset is $\{X: X_1 \in [\beta_2, 1], X_2 = 1\}$. Since $G(v)$ is identified for $v \in [\beta_2, 1]$ by observations of X_1 for which $X_2 = 0$, β_2 can be identified by solving

$$(2.15) \qquad E(Y|x_1,1) = G(x_1 + \beta_1) .$$

on the set of x_1 values where the ranges of $E(Y|x_1, 1)$ and $G(x_1 + \beta_2)$ overlap. ■

To see why G must satisfy a non-periodicity condition, suppose that in Example 2.3 G were periodic on $[\beta_2, 1]$ instead of strictly increasing. Then (2.15) would have at least two solutions, so β_2 would not be identified. The assumption that G is strictly increasing on [0,1] prevents this kind of periodicity, but many other shapes of G also satisfy the non-periodicity requirement. See Ichimura (1993) for details.

2.4.2 Identification Analysis when X Is Discrete

One of the conclusions reached in Section 2.4.1 is that β and G are not identified in a semiparametric single-index model if all components of X are discrete. It does not necessarily follow, however, that data are completely uninformative about β. In this section it is shown that if G is assumed to be an increasing function, then one can obtain identified *bounds* on the components of β.

To begin, it can be seen from Table 2.1 that there is a G that solves (2.14) for every possible value of β_2 in Example 2.3. Therefore, nothing can be learned about β_2 if nothing is known about G. This is not surprising. Even when the components of X are all continuous, some information about G is necessary to identify β (e.g., differentiability in the case of Theorem 2.1). Continuity and differentiability of G are not useful for identification when all components of X are discrete. A property that is useful, however, is monotonicity. The usefulness of this property is illustrated by the following example, which is a continuation of Example 2.3.

Example 2.5: Identification when X Is Discrete and G Is Monotonic

Consider the model of Example 2.3 and Table 2.1 but with the additional assumption that G is a strictly increasing function. That is,

$$(2.16) \qquad G(v_1) < G(v_2) \Leftrightarrow v_1 < v_2 .$$

Inequality (2.16) together with the information in columns 2 and 3 of Table 2.1 implies that $\beta_2 > 1$. This result is informative, even though it does not point identify β_2, because any value of β_2 in $(-\infty, \infty)$ is possible in principle. Knowledge of the population distribution of (Y, X) combined with monotonicity of G excludes all values in $(-\infty, 1]$.

If the support of X is large enough, then it is possible to identify an upper bound on β_2 as well as a lower bound. For example, suppose that the point $(X_1, X_2) = (0.6, 0.5)$ is in the support of X along with the four points in Example 2.3 and that $E(Y|X_1 = 0.6, X_2 = 0.5) = G(0.6 + 0.5\beta_2) = 0.35$. This information combined with (2.16) and row 3 of Table 2.1 implies that $\beta_2 < 0.6 + 0.5\beta_2$, so $\beta_2 < 1.2$. Therefore, the available information gives the identified bounds $1 < \beta_2 < 1.2$. Any value of β_2 within the interval (1, 1.2) is logically possible given the available information, so the bounds $1 < \beta_2 < 1.2$ are the tightest possible. ■

Now consider the general case in which X is k-dimensional for any finite $k \geq 2$ and has M points of support for any finite $M \geq 2$. Let x_m denote the m'th point of support ($m = 1,\ldots, M$). The population distribution of (Y, X) identifies $G(x_m\beta)$ for each m. Assume without loss of generality that the support points x_m are sorted so that

$$G(x_1\beta) \leq G(x_2\beta) \leq \ldots \leq G(x_M\beta) .$$

Achieve location and scale normalization by assuming that X has no constant component and that $\beta_1 = 1$. Also, assume that G is strictly increasing. Then tight, identified bounds on β_m ($2 \leq m \leq M$) can be obtained by solving the linear programming problems

$$(2.17) \qquad \begin{aligned} &\text{maximize (minimize): } b_m \\ &\text{subject to: } x_j b \leq x_{j+1} b; \quad j = 1,\ldots M-1 \end{aligned}$$

with strict equality holding in the constraint if $G(x_j b) = G(x_{j+1} b)$. The solutions to these problems are informative whenever they are not infinite.

Bounds on other functionals of β can be obtained by suitably modifying the objective function of (2.17). For example, suppose that x^* is a point that is not in the support of X and that we are interested in learning whether $E(Y|X = x^*) = G(x^*\beta)$ is larger or smaller than $E(Y|X=x_m) = G(x_m\beta)$ for some x_m in the support of X. $G(x^*\beta) - G(x_m\beta)$ is not identified if X is discrete, but $(x^* - x_m)\beta$ can be bounded by replacing b_m with $(x^* - x_m)b$ in the objective function of (2.17). If the resulting lower bound exceeds zero, then we know that $G(x^*\beta) > G(x_m\beta)$, even though $G(x^*\beta)$ is unknown. Similarly, $G(x^*\beta) < G(x_m\beta)$, if the upper bound obtained from the modified version of (2.17) is negative.

Now consider solving (2.17) with the objective function $(x_m - x^*)b$ for each $m = 1,\ldots, M$. Suppose this procedure yields the result $(x_m - x^*)\beta < 0$ if $m \le j$ for some j $(1 \le j \le M)$. Then it follows from monotonicity of G that $G(x^*\beta) > G(x_j\beta)$. Similarly, if the solutions to the modified version of (2.17) yield the result $(x_m - x^*)\beta > 0$ if $m \ge k$ for some k $(1 \le k \le M)$, then $G(x^*\beta) < G(x_k\beta)$. Since $G(x_j\beta)$ and $G(x_k\beta)$ are identified, this procedure yields identified bounds on the unidentified quantity $G(x^*\beta)$, thereby providing a form of extrapolation in a single-index model with a discrete X. The following example illustrates this form of extrapolation.

Example 2.6: Extrapolation when X Is Discrete and G is Monotonic

Let G, $E(Y|x)$, and the points of support of X be as in Example 2.5. Order the points of support as in Table 2.2. As in Example 2.5, the available information implies that

$$(2.18) \qquad 1 < \beta_2 < 1.2$$

but does not further identify β_2. Suppose that $x^* = (0.3, 0.25)$. What can be said about the value of $E(Y|x^*) = G(x^*\beta) = G(0.3 + 0.25\beta_2)$? This quantity is not identified, but the following bounds may be obtained by combining the information in Table 2.2 with inequality (2.18):

$$-0.6 < (x_1 - x^*)\beta < -0.55,$$

$$0.4 < (x_2 - x^*)\beta < 0.45,$$

$$0.45 < (x_3 - x^*)\beta < 0.60,$$

Table 2.2: A Second Unidentified Single-Index Model

m	x_m	$E(Y\|x_m)$	$G(x_m)$
1	(0, 0)	0	$G(0)$
2	(1, 0)	0.1	$G(1)$
3	(0, 1)	0.3	$G(\beta_2)$
4	(0.6, 0.5)	0.35	$G(0.6 + 0.5\beta_2)$
5	(1, 1)	0.4	$G(1 + \beta_2)$

$$0.55 < (x_4 - x^*)\beta < 0.60,$$

and

$$1.45 < (x_5 - x^*)\beta < 1.60.$$

Therefore, monotonicity of G implies that $G(x_1\beta) < G(x^*\beta) < G(x_2\beta)$, so identified bounds on the unidentified quantity $G(x^*\beta)$ are $0 < G(x^*\beta) < 0.1$. ■

2.5 Estimating G in a Single-Index Model

We now turn to the problem of estimating G and β in the single-index model (2.1). It is assumed throughout the remainder of this chapter that G and β are identified. This section is concerned with estimating G. Estimation of β is dealt with in Sections 2.6 and 2.7.

Suppose, for the moment, that β is known. Then G can be estimated as the nonparametric mean-regression of Y on $X\beta$. There are many nonparametric mean-regression estimators that can be used. See, for example Härdle (1990) and Härdle and Linton (1994). This book uses kernel estimators. The properties of these estimators are summarized in the Appendix.

To obtain a kernel estimator of $G(z)$ at any z in the support of $X\beta$, let the data consist of a random sample of n observations of (Y, X). Let $\{Y_i, X_i\colon i = 1, \ldots, n\}$ denote the sample. Here, the subscript i indexes observations, not components of X. Define $Z_i = X_i\beta$. Let K be a kernel function, and let $\{h_n\}$ be a sequence of bandwidth parameters. Under the assumption that β is known, the kernel nonparametric estimator of $G(z)$ is

$$(2.19) \qquad G_n^*(z) = \frac{1}{nh_n p_n^*(z)} \sum_{i=1}^{n} Y_i K\left(\frac{z - Z_i}{h_n}\right),$$

where

$$(2.20) \qquad p_n^*(z) = \frac{1}{nh_n} \sum_{i=1}^{n} K\left(\frac{z - Z_i}{h_n}\right).$$

The estimator (2.19) cannot be implemented in an application because β and, therefore, Z_i are not known. This problem can be remedied by replacing the unknown β with an estimator, b_n. Define $Z_{ni} = X_i b_n$ to be the corresponding estimator of Z_i. The resulting kernel estimator of G is

(2.21) $$G_n(z) = \frac{1}{nh_n p_n(z)} \sum_{i=1}^{n} Y_i K\left(\frac{z - Z_{ni}}{h_n}\right),$$

where

(2.22) $$p_n(z) = \frac{1}{nh_n} \sum_{i=1}^{n} K\left(\frac{z - Z_{ni}}{h_n}\right).$$

It is shown in Sections 2.6 and (2.7) that β can be estimated with a $n^{-1/2}$ rate of convergence in probability. That is, there exist estimators b_n with the property that $(b_n - \beta) = O_p(n^{-1/2})$. This is faster than the fastest possible rate of convergence in probability of a nonparametric estimator of $E(Y|z)$. As a result, the difference between the estimators G_n* and G_n is asymptotically negligible. Specifically,

$$(nh_n)^{1/2}[G_n(z) - G(z)] = (nh_n)^{1/2}[G_n^*(z) - G(z)] + o_p(1).$$

for any z in the support of Z. Therefore, estimation of β has no effect on the asymptotic distributional properties of the estimator of β. The reasoning behind this conclusion is easily outlined. Let, $\widetilde{b}_n$ and $\widetilde{\beta}$, respectively, denote the vectors obtained from b_n and β by removing their first components (the components set by scale normalization). Let $\widetilde{X}_i$ be the vector obtained from X_i, by removing its first component. Define K' to be the derivative of the kernel function K. For any $\widetilde{b}$ and $b \equiv (1, \widetilde{b}\,')'$, define

$$A_n(\widetilde{b}) = \frac{1}{nh_n} \sum_{i=1}^{n} Y_i K\left(\frac{z - X_i b}{h_n}\right),$$

$$A_{nz}(\widetilde{b}) = -\frac{1}{nh_n^2} \sum_{i=1}^{n} Y_i K'\left(\frac{z - X_i b}{h_n}\right)\widetilde{X}_i,$$

$$\widetilde{p}_n(\widetilde{b}) = \frac{1}{nh_n} \sum_{i=1}^{n} K\left(\frac{z - X_i b}{h_n}\right),$$

and

$$\tilde{p}_{nz}(\tilde{b}) = -\frac{1}{nh_n^2}\sum_{i=1}^{n} K'\left(\frac{z - X_i b}{h_n}\right)\tilde{X}_i \,.$$

Now make a Taylor series expansion of (2.21) about $b_n = \beta$ to obtain

$$(2.23) \qquad G_n(z) = G_n^*(z) + \frac{\partial G_n(z)}{\partial \tilde{b}_n^*}(\tilde{b}_n - \tilde{\beta}),$$

where $\tilde{b}_n^*$ is between $\tilde{b}_n$ and $\tilde{\beta}$ and for any $\tilde{b}$

$$(2.24) \qquad \frac{\partial G_n(z)}{\partial \tilde{b}} = \frac{A_{nz}(\tilde{b})}{\tilde{p}_n(\tilde{b})} - \frac{A_n(\tilde{b})\tilde{p}_{nz}(\tilde{b})}{\tilde{p}_n^2(\tilde{b})}.$$

By using a suitable uniform law of large numbers (see, e.g., Pollard 1984), it can be shown that each term on the right-hand side of (2.24) converges in probability to a nonstochastic limit. Therefore, there is a nonstochastic function Γ such that

$$(2.25) \qquad \frac{\partial G_n(z)}{\partial \tilde{b}_n^*} = \Gamma(z) + o_p(1).$$

It follows from (2.23), (2.25), and $(b_n - \beta) = O_p(n^{-1/2})$ that

$$G_n(z) - G_n^*(z) = \Gamma(z)(\tilde{b}_n - \tilde{\beta}) + o_p[(\tilde{b}_n - \tilde{\beta})]$$

$$= O_p(n^{-1/2}).$$

This implies that

$$(2.26) \qquad (nh_n)^{1/2}[G_n(z) - G_n^*(z)] = O_p(h_n^{1/2}),$$

which gives the desired result

The foregoing results concerning estimation of G apply with any b_n that is a $n^{1/2}$-consistent estimator of β. We now turn to developing such estimators.

2.6 Optimization Estimators of β

Estimators of β can be classified according to whether they require solving nonlinear optimization problems. This section discusses estimators that are obtained as the solutions to nonlinear optimization problems. Section 2.7 discusses estimators that do not require solving optimization problems.

2.6.1 Nonlinear Least Squares

If G were known, then β could be estimated by nonlinear least squares or weighted nonlinear least squares (WNLS). Let the data consist of the random sample $\{Y_i, X_i: \ i = 1,\ldots, n\}$. Then the WNLS estimator of β, b_{NLS}, is the solution to

$$(2.27) \quad \text{minimize:} \ S_n^*(b) = \frac{1}{n}\sum_{i=1}^{n} W(X_i)[Y_i - G(X_i b)]^2$$

where W is the weight function. Under mild regularity conditions, b_{NLS} is a consistent estimator of β, and $n^{1/2}(b_{NLS} - \beta)$ is asymptotically normally distributed with a mean of zero and a covariance matrix that can be estimated consistently. See, for example, Amemiya (1985), Davidson and MacKinnon (1993), and Gallant (1987).

The estimator b_{NLS} is not available in the semiparametric case, where G is unknown. Ichimura (1993) showed that this problem can be overcome by replacing the G in (2.27) with a suitable estimator. This estimator is a modified version of the kernel estimator (2.21). Three modifications are needed. First, observe that if G_n is defined as in (2.21), then the denominator of $G_n(X_i b)$ contains the term $p_n(X_i b)$. To keep this term from getting arbitrarily close to zero as n increases, it is necessary to restrict the sums in (2.21) and (2.27) to observations i for which the probability density of $X\beta$ at the point $X_i\beta$ exceeds a small, positive number. Second, it is necessary to exclude observation i from the calculation of $G_n(X_i b)$. Third, it is necessary to weight the terms of the sums in the calculation of G_n the same way that the terms in the sum (2.27) are weighted.

To carry out these modifications, let $p(\bullet, b)$ denote the probability density function of Xb. Let B be a compact set that contains β. Define A_x and A_{nx} to be the following sets:

$$A_x = \{x: p(xb, b) \geq \eta \ \ \forall b \in B\},$$

and

$$A_{nx} = \{x: \|x - x^*\| \le 2h_n \text{ for some } x^* \in A_x\},$$

where $\eta > 0$ is a constant, h_n is the bandwidth used for kernel estimation, and $\|\bullet\|$ is the Euclidean norm. Let $I(\bullet)$ denote the indicator function. $I(\bullet) = 1$ if the event in parentheses occurs and 0 otherwise. Define $J_j = I(X_j \in A_x)$ and $J_{nj} = I(X_j \in A_{nx})$. Finally, define

$$(2.28) \qquad G_{ni}(z,b) = \frac{1}{nh_n p_{ni}(z,b)} \sum_{j \ne i} Y_j J_{nj} W(X_j) K\left(\frac{z - X_j b}{h_n}\right),$$

where for any z

$$(2.29) \qquad p_{ni}(z,b) = \frac{1}{nh_n} \sum_{j \ne i} J_{nj} W(X_j) K\left(\frac{z - X_j b}{h_n}\right).$$

The estimator of $G(X_i b)$ that is used in (2.27) is $G_{ni}(X_i b)$. Thus, the semiparametric WNLS estimator of β is the solution to

$$(2.30) \qquad \text{minimize } S_n(\tilde{b}) = \frac{1}{n} \sum_{i=1}^{n} J_i W(X_i)[Y_i - G_{ni}(X_i b, b)]^2.$$

The minimization is over $\tilde{b}$, not b, to impose scale normalization. Let $\tilde{b}_n$ denote the resulting estimator, and call it the semiparametric WNLS estimator of $\tilde{\beta}$.

Ichimura (1993) gives conditions under which $\tilde{b}_n$ is a consistent estimator of $\tilde{\beta}$ and

$$(2.31) \qquad n^{1/2}(\tilde{b}_n - \tilde{\beta}) \to^d N(0, \Omega).$$

The covariance matrix, Ω, is given in equation (2.32) below. The conditions under which (2.31) holds are stated in Theorem 2.2.

Theorem 2.2: Equation (2.31) holds if the following conditions are satisfied:

(a) $\{Y_i, X_i: i = 1,\ldots,n\}$ is a random sample from a distribution that satisfies (2.1).

(b) β is identified and is an interior point of the known compact set, B.

(c) A_x is compact, and W is bounded and positive on A_x.

(d) $E(Y|Xb = z)$ and $p(z,b)$ are 3 times continuously differentiable with respect to z. The third derivatives are Lipschitz continuous uniformly over B for all $z \in \{z: z = xb, b \in B, x \in A_x\}$.

(e) $E|Y^m| < \infty$ for some $m \geq 3$. The variance of Y conditional on $X = x$ is bounded and bounded away from 0 for $x \in A_x$.

(f) The kernel function K is twice continuously differentiable, and its second derivative is Lipschitz continuous. Moreover $K(v) = 0$ if $|v| > 1$, and

$$\int v^j K(v)dv = \begin{cases} 1 & \text{if } j = 0 \\ 0 & \text{if } j = 1 \end{cases}$$

(g) The bandwidth sequence $\{h_n\}$ satisfies $(\log h_n)/[nh_n^{3+3/(m-1)}] \to 0$ and $nh_n^8 \to 0$ as $n \to \infty$. ■

There are several noteworthy features of Theorem 2.2. First, $\tilde{b}_n$ converges in probability to $\tilde{\beta}$ at the rate $n^{-1/2}$, which is the same rate that would be obtained if G were known and faster than the rate of convergence of a nonparametric density or mean-regression estimator. This result was used in deriving (2.26). Second, the asymptotic distribution of $n^{1/2}(\tilde{b}_n - \tilde{\beta})$ is centered at zero. This contrasts with the case of nonparametric density and mean-regression estimators, whose asymptotic distributions are not centered at zero in general when the estimators have their fastest possible rates of convergence. Third, the range of permissible rates of convergence of h_n includes the rate $n^{-1/5}$, which is the standard rate in nonparametric density and mean-regression estimation. Finally, Theorem 2.2 requires β to be contained in the known, compact set B. Therefore, in principle $S_n(\tilde{b})$ should be minimized subject to the constraint $\tilde{b} \in B$. In practice, however, the probability that the constraint is binding for any reasonable B is so small that it can be ignored. This is a useful result because solving a constrained nonlinear optimization problem is usually much more difficult than solving an unconstrained one.

Stating the covariance matrix, Ω, requires additional notation. Let $p(\bullet|\tilde{x},b)$ denote the probability density function of Xb conditional on $\tilde{X} = \tilde{x}$. Define $p(\bullet|\tilde{x}) = p(\bullet|\tilde{x},\beta)$, $\sigma^2(x) = \text{Var}(Y|X = x)$, and

$$G(z,b) = \operatorname*{plim}_{n\to\infty} G_{ni}(z,b).$$

Calculations that are lengthy but standard in kernel estimation show that

$$G(z,b) = \frac{E[E(Y|z,\widetilde{X})I(X \in A_x)W(X)p(z|\widetilde{X},b)]}{E[I(X \in A_x)W(X)p(z|\widetilde{X},b)]}$$

$$= \frac{R_1(z,b)}{R_2(z,b)},$$

where

$$R_1(z,b) = E\{G[z - \widetilde{X}(\widetilde{b} - \widetilde{\beta})]$$

$$\bullet\, p[z - \widetilde{X}(\widetilde{b} - \widetilde{\beta})|\widetilde{X}]W(X)I(X \in A_x)\}$$

and

$$R_2(z,b) = E\{p[z - \widetilde{X}(\widetilde{b} - \widetilde{\beta})|\widetilde{X}]$$

$$\bullet\, W(X)I(X \in A_x)\}.$$

Moreover,

$$G(z,\beta) = G(z)$$

and for $z = x\beta$

$$\frac{\partial G(z,\beta)}{\partial \widetilde{b}} = G'(z)\left\{\widetilde{x} - \frac{E[\widetilde{X}W(X)|X\beta = z, X \in A_x]}{E[W(X)|X\beta = z, X \in A_x]}\right\}. \tag{2.32}$$

Now define

$$C = E\left[I(X \in A_x)W(X)\frac{\partial G(X\beta,\beta)}{\partial \widetilde{b}}\frac{\partial G(X\beta,\beta)}{\partial \widetilde{b}'}\right],$$

and

$$D = E\left[I(X \in A_x)W^2(X)\sigma^2(X)\frac{\partial G(X\beta,\beta)}{\partial \widetilde{b}}\frac{\partial G(X\beta,\beta)}{\partial \widetilde{b}'}\right].$$

Then

(2.33) $\quad \Omega = C^{-1} D C^{-1}.$

Theorem 2.2 is proved in Ichimura (1993). The technical details of the proof are complex, but the main ideas are straightforward and based on the familiar Taylor series methods of asymptotic distribution theory. With probability approaching one as $n \to \infty$, the solution to (2.30) satisfies the first-order condition

$$\frac{\partial S_n(\tilde{b}_n)}{\partial \tilde{b}} = 0.$$

Therefore, a Taylor series expansion gives

(2.34) $$n^{1/2} \frac{\partial S_n(\tilde{\beta})}{\partial \tilde{b}} = \frac{\partial^2 S_n(\bar{b}_n)}{\partial \tilde{b}\, \partial \tilde{b}'} n^{1/2} (\tilde{b}_n - \tilde{\beta}),$$

where $\bar{b}_n$ is between $\tilde{b}_n$ and $\tilde{\beta}$. Now consider the left-hand side of (2.34). Differentiation of S_n gives

$$n^{1/2} \frac{\partial S_n(\tilde{\beta})}{\partial \tilde{b}} =$$

$$-\frac{1}{n^{1/2}} \sum_{i=1}^{n} J_i W(X_i)[Y_i - G_{ni}(X_i\beta, \beta)] \frac{\partial G_{ni}(X_i\beta, \beta)}{\partial \tilde{b}}.$$

Moreover,

$$G_{ni}(X_i\beta, \beta) \to^p G(X_i\beta)$$

and

$$\frac{\partial G_{ni}(X_i\beta, \beta)}{\partial \tilde{b}} \to^p \frac{\partial G(X_i\beta, \beta)}{\partial \tilde{b}}$$

sufficiently rapidly that we may write

(2.35) $$n^{1/2}\frac{\partial S_n(\tilde{\beta})}{\partial \tilde{b}} =$$

$$-\frac{1}{n^{1/2}}\sum_{i=1}^{n} J_i W(X_i)[Y_i - G(X_i\beta)]\frac{\partial G(X_i\beta,\beta)}{\partial \tilde{b}} + o_p(1).$$

The first term on the right-hand side of (2.35) is asymptotically distributed as $N(0,D)$ by the multivariate generalization of the Lindeberg-Levy central limit theorem. Therefore, the left-hand side of (2.34) is also asymptotically distributed as $N(0,D)$.

Now consider the right-hand side of (2.34). Differentiation of S_n gives

$$\frac{\partial^2 S_n(\bar{b}_n)}{\partial \tilde{b}\,\partial \tilde{b}'} = \frac{1}{n}\sum_{i=1}^{n} J_i W(X_i)\frac{\partial G_{ni}(X_i\bar{b}_n',\bar{b}_n')}{\partial \tilde{b}}\frac{\partial G_{ni}(X_i\bar{b}_n,\bar{b}_n)}{\partial \tilde{b}'}$$

$$-\frac{1}{n}\sum_{i=1}^{n} J_i W(X_i)[Y_i - G_{ni}(X_i\bar{b}_n,\bar{b}_n)]\frac{\partial^2 G_{ni}(X_i\bar{b}_n',\bar{b}_n)}{\partial \tilde{b}\,\partial \tilde{b}'}.$$

Because $G_{ni}(xb,b)$ and its derivatives converge to $G(xb,b)$ and its derivatives uniformly over both arguments, we may write

$$\frac{\partial^2 S_n(\bar{b}_n)}{\partial \tilde{b}\,\partial \tilde{b}'} =$$

$$+\frac{1}{n}\sum_{i=1}^{n} J_i W(X_i)\frac{\partial G(X_i\beta,\beta)}{\partial \tilde{b}}\frac{\partial G(X_i\beta,\beta)}{\partial \tilde{b}'}$$

$$-\frac{1}{n}\sum_{i=1}^{n} J_i W(X_i)[Y_i - G(X_i\beta,\beta)]\frac{\partial^2 G(X_i\beta,\beta)}{\partial \tilde{b}\,\partial \tilde{b}'}$$

$$+o_p(1).$$

The first term on the right-hand side of this equation converges almost surely to C and the second term converges to almost surely to zero by the strong law of large numbers. This result together with the previously obtained asymptotic distribution of the left-hand side of (2.34) implies that (2.34) can be written in the form

$$(2.36) \qquad N(0,D) = Cn^{1/2}(\tilde{b}_n - \beta) + o_p(1).$$

Equation (2.13) is obtained by multiplying both sides of (2.36) by C^{-1}.

In applications, Ω is unknown, and a consistent estimator is needed to make statistical inference possible. To this end, define.

$$C_n = \frac{1}{n}\sum_{i=1}^{n} J_i W(X_i) \frac{\partial G_{ni}(X_i b_n, b_n)}{\partial \tilde{b}} \frac{\partial G_{ni}(X_i b_n, b_n)}{\partial \tilde{b}'}$$

and

$$D_n =$$

$$\frac{1}{n}\sum_{i=1}^{n} J_i W(X_i)[Y_i - G_{ni}(X_i b_n)]^2 \frac{\partial G_{ni}(X_i b_n, b_n)}{\partial \tilde{b}} \frac{\partial G_{ni}(X_i b_n, b_n)}{\partial \tilde{b}'}.$$

Under the assumptions of Theorem 2.2, C_n and D_n, respectively, are consistent estimators of C and D. Ω is estimated consistently by

$$\Omega_n = C_n^{-1} D_n C_n^{-1}.$$

Intuitively, these results can be understood by observing that because G_{ni} converges in probability to G and b_n converges in probability to β,

$$C_n = \frac{1}{n}\sum_{i=1}^{n} J_i W(X_i) \frac{\partial G(X_i \beta, \beta)}{\partial \tilde{b}} \frac{\partial G(X_i \beta, \beta)}{\partial \tilde{b}'} + o_p(1)$$

and

$$D_n =$$

$$\frac{1}{n}\sum_{i=1}^{n} J_i W(X_i)[Y_i - G(X_i \beta)]^2 \frac{\partial G(X_i \beta, \beta)}{\partial \tilde{b}} \frac{\partial G(X_i \beta, \beta)}{\partial \tilde{b}'} + o_p(1).$$

Convergence of C_n to C and D_n to D now follows from the strong law of large numbers.

2.6.2 Choosing the Weight Function

The choice of weight function, W, affects the efficiency of the estimator of $\tilde{\beta}$. Ideally, one would like to choose W so as to maximize the asymptotic efficiency of the estimator. Some care is needed in defining the concept of asymptotic efficiency so as to avoid the pathology of superefficiency. See Bickel *et al.* (1993) and Ibragimov and Has'minskii (1981) for discussions of superefficiency and methods for avoiding it. Estimators that are restricted so as to avoid superefficiency are called *regular*.

Within the class of semiparametric WNLS estimators, an estimator is asymptotically efficient if the covariance matrix Ω of its asymptotic distribution differs from the covariance matrix Ω^* of any of any other weighted WNLS estimator by a positive semidefinite matrix. That is, $\Omega^* - \Omega$ is positive semidefinite. More generally, one can consider the class of all regular estimators of single-index models (2.1). This class includes estimators that may not be semiparametric WNLS estimators. The definition of an asymptotically efficient estimator remains the same, however. The covariance matrix of the asymptotic distribution of any regular estimator exceeds that of the asymptotically efficient estimator by a positive semidefinite matrix.

The problem of asymptotically efficient estimation of β in a semiparametric single-index model is related to but more difficult than the problem of asymptotically efficient estimation in a nonlinear regression model with a known G. The case of a nonlinear regression model (not necessarily a single-index model) in which G is known has been investigated by Chamberlain (1987), who derived an asymptotic efficiency bound. The covariance matrix of the asymptotic distribution of any regular estimator must exceed this bound by a positive semidefinite matrix. The model is $E(Y|x) = G(x,\beta)$. The variance function, $\sigma^2(x) = E\{[Y - G(X,\beta)]^2|X = x\}$, is unknown. Chamberlain (1986) showed that the efficiency bound is

$$\Omega_{NLR} = \left\{ E\left[\sigma^{-2}(X) \frac{\partial G(X,\beta)}{\partial b} \frac{\partial G(X,\beta)}{\partial b'} \right] \right\}^{-1}.$$

This is the covariance matrix of a weighted (or generalized) nonlinear least squares estimator of β with weight function $W(x) = \sigma^{-2}(x)$. For the special case of a the linear model $G(x,\beta) = x\beta$, Carroll (1982) and Robinson (1987) showed that this covariance matrix is obtained asymptotically even when $\sigma^2(x)$ is unknown by replacing $\sigma^2(x)$ with a nonparametric estimator. Thus, lack of knowledge of $\sigma^2(x)$ causes no loss of asymptotic efficiency relative to infeasible generalized least squares estimation.

The problem of efficient estimation of β in a single-index model with an unknown G has been investigated by Ichimura and Hall (1991) and Newey and

Stoker (1993). These authors showed that under regularity conditions, the efficiency bound for estimating β in a single-index model with unknown G and using only data for which $X \in A_x$ is (2.33) with weight function $W(x) = \sigma^2(x)$. With this weight function, $C = D$ in (2.33), so the efficiency bound is

$$(2.37) \qquad \Omega_{SI} \equiv \left\{ E\left[\frac{I(X \in A_x)}{\sigma^2(X)} \frac{\partial G(X\beta, \beta)}{\partial \tilde{b}} \frac{\partial G(X\beta, \beta)}{\partial \tilde{b}'} \right] \right\}^{-1}.$$

This bound is achieved by the semiparametric WNLS estimator if $\sigma^2(x)$ is known. The assumption that the estimator uses only observations for which $X \in A_x$ can be eliminated by letting A_x grow very slowly as n increases. Chamberlain (1986) and Cosslett (1987) derived this asymptotic efficiency bound for the case in which (2.1) is a binary response model (that is, the only possible values of Y are 0 and 1) and G is a distribution function. Chamberlain and Cosslett also derived efficiency bounds for certain kinds of censored regression models. Except in special cases, Ω_{SI} exceeds the asymptotic efficiency bound that would be achievable if G were known. Thus, there is a cost in terms of asymptotic efficiency (but not rate of convergence of the estimator) for not knowing G. Cosslett (1987) gives formulae for the efficiency losses in binary-response and censored linear regression models.

When $\sigma^2(x)$ is unknown, as is likely in applications, it can be replaced by a consistent estimator. Call this estimator $s_n^2(x)$. The asymptotic efficiency bound will be achieved by setting $W(x) = 1/s_n^2(x)$ in the semiparametric WNLS estimator (Newey and Stoker, 1993). Therefore, an asymptotically efficient estimator of β can be obtained even when $\sigma^2(x)$ is unknown.

A consistent estimator of $\sigma^2(x)$ can be obtained by using the following two-step procedure. In the first step, estimate β by using semiparametric WNLS with $W(x) = 1$. The resulting estimator is $n^{1/2}$- consistent and asymptotically normal but inefficient. Let e_i be the i'th residual from the estimated model. That is, $e_i = Y_i - G_{ni}(X_i b_n, b_n)$. In the second step, set $s_n^2(x)$ equal to a nonparametric estimator of the mean regression of e_i^2 on X_i. Robinson (1987) discusses technical problems that arise if X has unbounded support or a density that can be arbitrarily close to zero. He avoids these problems by using a nearest-neighbor nonparametric regression estimator. In practice, a kernel estimator will suffice if A_x is chosen so as to keep the estimated density of X away from zero.

This concludes the discussion of semiparametric weighted nonlinear least squares estimation of single-index models. To summarize, Ichimura (1993) has given conditions under which the semiparametric WNLS estimator of β in (2.1) is $n^{1/2}$-consistent and asymptotically normal. The estimator of β is also asymptotically efficient if the weight function is a consistent estimator of $\sigma^2(x)$. A consistent estimator of $\sigma^2(x)$ can be obtained by a two-step procedure in which the first step is semiparametric WNLS estimation of β with a unit weight function, and the second step is nonparametric estimation of the mean of the squared first-step residuals conditional on X.

2.6.3 Semiparametric Maximum-Likelihood Estimation of Binary-Response Models

This section is concerned with estimation of (2.1) when the only possible values of Y are 0 and 1. In this case, $G(x\beta) = P(Y=1|x)$. If G were a known function, then the asymptotically efficient estimator of β would be the maximum-likelihood estimator (MLE). The MLE solves the problem

$$\text{maximize:} \quad \log[L(b)] =$$

(2.38)

$$n^{-1}\sum_{i=1}^{n}\{Y_i \log[G(X_i b)] + (1-Y_i)\log[1-G(X_i b)]\}.$$

In the semiparametric case, where G is unknown, one can consider replacing G on the right-hand side of (2.38) with an estimator such as G_{ni} in (2.28). This idea has been investigated in detail by Klein and Spady (1993). It is clear from (2.38) that care must be taken to insure that any estimate of G is kept sufficiently far from 0 and 1. Klein and Spady (1993) use elaborate trimming procedures to accomplish this without artificially restricting X to a fixed set A_x on which $G(X\beta)$ is bounded away from 0 and 1. They find, however, that trimming has little effect on the numerical performance of the resulting estimator. Therefore, in practice little is lost in terms of estimation efficiency and much is gained in simplicity by using only observations for which $X \in A_x$. This method will be used in the remainder of this section.

A second simplification can be obtained by observing that in the special case of a binary-response model, $\mathrm{Var}(Y|x) = G(x\beta)[1 - G(x\beta)]$. Thus, $\sigma^2(x)$ depends only on the index $z = x\beta$. In this case, W cancels out of the numerator and denominator terms on the right-hand side of (2.32), so

$$\frac{\partial G(z,\beta)}{\partial \tilde{b}} = G'(z)\left\{\tilde{x} - E[\tilde{X}|X\beta = z, X \in A_x]\right\}.$$

By substituting this result into (2.33) and (2.37), it can be seen that the covariance matrix of the asymptotic distribution of the semiparametric WNLS estimator of β is the same whether the estimator of G is weighted or not. Moreover, the asymptotic efficiency bound Ω_{SI} can be achieved without weighting the estimator of G. Accordingly, define the unweighted estimator of G

$$\hat{G}_{ni}(z,b) = \frac{1}{nh_n \hat{p}_{ni}(z,b)} \sum_{j \neq i} Y_j J_{nj} K\left(\frac{z - X_j b}{h_n}\right),$$

where

$$\hat{p}_{ni}(z,b) = \frac{1}{nh_n} \sum_{j \neq i} J_{nj} K\left(\frac{z - X_j b}{h_n}\right).$$

Now consider the following semiparametric analog of (2.38):

(2.39) maximize: $\log[L_{SP}(\tilde{b})] =$

$$n^{-1} \sum_{i=1}^{n} J_i \{Y_i \log[\hat{G}_{ni}(X_i b, b)] + (1 - Y_i) \log[1 - \hat{G}_{ni}(X_i b, b)]\}.$$

Let $\hat{b}_n$ denote the resulting estimator of $\tilde{\beta}$. If β is identified (see the discussion in Section 2.4), consistency of $\hat{b}_n$ for $\tilde{\beta}$ can be demonstrated by showing that $\hat{G}_{ni}(z,b)$ converges to $G(z,b)$ uniformly over z and b. Therefore, the probability limit of the solution to (2.39) is the same as the probability limit of the solution to

(2.40) maximize: $\log[L_{SP}{}^*(\tilde{b})] =$

$$n^{-1} \sum_{i=1}^{n} J_i \{Y_i \log[G(X_i b, b)] + (1 - Y_i) \log[1 - G(X_i b, b)]\}.$$

The solution to (2.39) is consistent for $\tilde{\beta}$ if the solution to (2.40) is. The solution to (2.40) is a parametric maximum-likelihood estimator. Consistency for $\tilde{\beta}$ can be proved using standard methods for parametric maximum-likelihood estimators. See, for example, Amemiya (1985).

By differentiating the right-hand side of (2.39), it can be seen that $b_n \equiv (1, \hat{b}_n{}')'$ satisfies the first-order condition

$$\frac{1}{n} \sum_{i=1}^{n} J_i \frac{Y_i - \hat{G}_{ni}(X_i b_n, b_n)}{\hat{G}_{ni}(X_i b_n, b_n)[1 - \hat{G}_{ni}(X_i b_n, b_n)]} \frac{\partial \hat{G}_{ni}(X_i b_n, b_n)}{\partial \tilde{b}} = 0$$

with probability approaching 1 as $n \to \infty$. This is the same as the first-order condition for semiparametric WNLS estimation of β with the estimated weight function

$$W(x) = \left\{\hat{G}_{ni}(xb_n, b_n)[1 - \hat{G}_{ni}(xb_n, b_n)]\right\}^{-1}$$

$$= \left\{G(x\beta)[1 - G(x\beta)]\right\}^{-1} + o_p(1)$$

$$= [\mathrm{Var}(Y|x)]^{-1} + o_p(1).$$

It now follows from the discussion of asymptotic efficiency in semiparametric WNLS estimation (Section 2.6.2) that the semiparametric maximum-likelihood estimator of β in a single-index binary-response model achieves the asymptotic efficiency bound Ω_{SI}.

The conclusions of this section may be summarized as follows. The semiparametric maximum-likelihood estimator of β in a single-index binary-response model solves (2.39). The estimator is asymptotically efficient and satisfies

$$n^{1/2}(\hat{b}_n - \tilde{\beta}) \to^d N(0, \Omega_{SI}).$$

2.6.4 Semiparametric Maximum Likelihood Estimation of Other Single-Index Models

Ai (1997) has extended semiparametric maximum likelihood estimation to single-index models other than binary-response models. As in the binary-response estimator of Klein and Spady (1993), Ai (1997) forms a quasi-likelihood function by replacing the unknown probability density function of the dependent variable conditional on the index with a nonparametric estimator. To illustrate, suppose that the probability distribution of the dependent variable Y depends on the explanatory variables X only through the index $X\beta$. Let $f(\bullet|v, \beta)$ denote the probability density function of Y conditional on $X\beta = v$. If f were known, then β could be estimated by parametric maximum likelihood. For the semiparametric case, in which f is unknown, Ai replaces f with a kernel estimator of the density of Y conditional on the index. He then maximizes a trimmed version of the resulting quasi-likelihood function. Under suitable conditions, the resulting semiparametric estimator of β is asymptotically efficient (in the sense of achieving the semiparametric efficiency bound). See Ai (1997) for the details of the trimming procedure and regularity conditions.

2.7 Direct Semiparametric Estimators

Semiparametric weighted nonlinear least squares and maximum-likelihood estimators have the significant practical disadvantage of being very difficult to compute. This is because they are solutions to nonlinear optimization problems whose objective functions may be non-convex (non-concave in the case of the maximum-likelihood estimator) or even multimodal. Moreover, computing the objective functions requires estimating a nonparametric mean-regression at each data point and, therefore, can be very slow.

This section describes an estimation approach that does not require solving an optimization problem and is non-iterative (hence the name *direct*). Direct estimates can be computed very quickly. Although direct estimators are not asymptotically efficient, an asymptotically efficient estimator can be obtained from a direct estimator in one additional, non-iterative, computational step. The relative computational simplicity of direct estimators makes them highly attractive for practical data analysis.

Section 2.7.1 describes direct estimation under the simplifying but often unrealistic assumption that X is a continuously distributed random vector. Section 2.7.2 shows how the direct estimation method can be extended to models in which some components of X are discrete. Section 2.7.3 describes the one-step method for obtaining an asymptotically efficient estimator from a direct estimate.

2.7.1 Average-Derivative Estimators

The idea underlying direct estimation of a single-index model when X is a continuously distributed random vector is very simple. Let (2.1) hold. Assume that G is differentiable, as is required for identification of β. Then

$$\text{(2.41)} \qquad \frac{\partial E(Y|x)}{\partial x} = \beta G'(x\beta)\,.$$

Moreover, for any bounded, continuous function W,

$$\text{(2.42)} \qquad E\left[W(X)\frac{\partial E(Y|X)}{\partial x}\right] = \beta E\left[W(X)G'(X\beta)\right].$$

The quantity on the left-hand side of (2.42) is called a *weighted average derivative* of $E(Y|x)$ with weight function W. Equation (2.42) shows that a weighted average derivative of $E(Y|x)$ is proportional to β. Owing to the need for scale normalization, β is identified only up to scale, so any weighted average derivative of $E(Y|x)$ is observationally equivalent to β. Thus, to estimate β, it suffices to estimate the left-hand side of (2.42) for some W. The scale normalization $\beta_1 = 1$ can be imposed, if desired, by dividing each component of the left-hand side of (2.42) by the first component.

The left-hand side of (2.42) can be estimated by replacing $\partial E(Y|X)/\partial x$ with a kernel (or other nonparametric) estimator and the population expectation $E[\bullet]$ with a sample average. Härdle and Stoker (1989), Powell, *et al.* (1989) and Stoker (1986, 1991a, 1991b) describe various ways of doing this. The discussion here concentrates on the method of Powell, *et al.* (1989), which is especially easy to analyze and implement.

To describe this method, let $p(\bullet)$ denote the probability density function of X, and set $W(x) = p(x)$. Then the left-hand side of (2.42) can be written in the form

$$E\left[W(X)\frac{\partial E(Y|X)}{\partial x}\right] = E\left[p(X)\frac{\partial E(Y|X)}{\partial x}\right]$$

$$= \int \frac{\partial E(Y|x)}{\partial x} p(x)^2 dx.$$

Assume that $p(x) = 0$ if x is on the boundary of the support of X. Then integration by parts gives

$$E\left[W(X)\frac{\partial E(Y|X)}{\partial x}\right] = -2\int E(Y|x)\frac{\partial p(x)}{\partial x}p(x)dx$$

$$= -2E\left[Y\frac{\partial p(X)}{\partial x}\right].$$

Define $\delta = E[W(X)\partial E(Y|X)/\partial x]$. Then δ is observationally equivalent to β up to scale normalization and, if $W(x) = p(x)$,

$$\delta = -2E\left[Y\frac{\partial p(X)}{\partial x}\right]. \tag{2.43}$$

A consistent estimator of δ can be obtained by replacing p with a nonparametric estimator and the expectation operator with a sample average. Let $\{Y_i, X_i: i = 1,\ldots, n\}$ denote the sample. The estimator of δ is

$$\delta_n = -2\sum_{i=1}^{n} Y_i \frac{\partial p_{ni}(X_i)}{\partial x}, \tag{2.44}$$

where $p_{ni}(X_i)$ is the estimator of $p(X_i)$. The quantity δ_n is called a *density-weighted average-derivative estimator.*

To implement (2.44), the estimator of p must be specified. A kernel estimator is attractive because it is relatively easily analyzed and implemented. To this end, let $k = \dim(X)$, and let K be a kernel function with a k-dimensional argument. Conditions that K must satisfy are given in Theorem 2.3 below. Let $\{h_n\}$ be a sequence of bandwidth parameters. Set

$$p_{ni}(x) = \frac{1}{n-1} \sum_{j \neq i} \left(\frac{1}{h_n} \right)^k K\left(\frac{x - X_j}{h_n} \right)$$

It follows from standard properties of kernel density estimators (see the Appendix) that $p_{ni}(x)$ is a consistent estimator of $p(x)$. Moreover, $\partial p(x)/\partial x$ is estimated consistently by $\partial p_{ni}(x)/\partial x$. The formula for $\partial p_{ni}(x)/\partial x$ is

$$\text{(2.45)} \qquad \frac{\partial p_{ni}(x)}{\partial x} = \frac{1}{n-1} \sum_{j \neq i} \left(\frac{1}{h_n} \right)^{k+1} K'\left(\frac{x - X_j}{h_n} \right),$$

where K' denotes the gradient of K. Substituting (2.45) into (2.44) yields

$$\text{(2.46)} \qquad \delta_n = -\frac{2}{n(n-1)} \sum_{i=1}^{n} \sum_{j \neq i} \left(\frac{1}{h_n} \right)^{k+1} K'\left(\frac{X_i - X_j}{h_n} \right) Y_i .$$

Observe that the right-hand side of (2.46) does not have a density estimator or other random variable in its denominator. This is because setting $W(x) = p(x)$ in the weighted average derivative defined in (2.42) cancels the density function that would otherwise be in the denominator of the estimator of $E(Y|x)$. This lack of a random denominator is the main reason for the relative ease with which δ_n can be analyzed and implemented.

Powell, *et al.* (1989) give conditions under which δ_n is a consistent estimator of δ and $n^{1/2}(\delta_n - \delta)$ is asymptotically normally distributed with mean 0. The formal statement of this result and the conditions under which it holds are stated in Theorem 2.3. Let $\|\bullet\|$ denote the Euclidean norm. Let $P = (k + 2)/2$ if k is even and $P = (k + 3)/2$ if k is odd.

Theorem 2.3: Let the following conditions hold.

(a) The support of X is a convex, possibly unbounded, subset of $\Re^k$ with a non-empty interior. X has a probability density function p. All partial derivatives of p up to order $P + 1$ exist.

(b) The components of $\partial E(Y|X)/\partial x$ and of the matrix $[\partial p(X)/\partial x][Y, X']$ have finite second moments. $E[Y\partial^r p(X)]$ exists for all positive integers $r \leq P + 1$, where $\partial^r p(x)$ denotes any order r mixed partial derivative of p. $E(Y^2|x)$ is a continuous function of x. There is a function $m(x)$ such that

$$E[(1+|Y|+\|X\|)m(X)]^2 < \infty ,$$

$$\left\| \frac{\partial p(x+\xi)}{\partial x} - \frac{\partial p(x)}{\partial x} \right\| < m(x)\|\xi\| ,$$

and

$$\left\| \frac{\partial[p(x+\xi)E(Y|x+\xi)]}{\partial x} - \frac{\partial[p(x)E(Y|x)]}{\partial x} \right\| < m(x)\|\xi\|.$$

(c) The kernel function K is symmetrical about the origin, bounded, and differentiable. The moments of K through order P are finite. The moments of K of order r are all 0 if $1 \leq r < P$. In addition

$$\int K(v)dv = 1..$$

(d) The bandwidth sequence $\{h_n\}$ satisfies $nh_n^{2P} \to 0$ and $nh_n^{k+2} \to \infty$ as $n \to \infty$.

Then

$$n^{1/2}(\delta_n - \delta) \to^d N(0, \Omega_{AD}) ,$$

where

$$\Omega_{AD} = 4E[R(Y,X)R(Y,X)'] - 4\delta\delta' \tag{2.47}$$

and

$$R(y,x) = p(x)\frac{\partial E(Y|x)}{\partial x} - [y - E(Y|x)]\frac{\partial p(x)}{\partial x} . \quad \blacksquare$$

A consistent estimator of Ω_{AD} is given in equation (2.54) below.

Several comments may be made about the conditions imposed in Theorem 2.3. Condition (a) implies that X is a continuously distributed random variable and that no component of X is functionally determined by other components. Condition (b) requires the existence of various moments and imposes smoothness requirements on $p(x)$, $E(Y|x)$, and $E(Y^2|x)$. Condition (c) requires K to be a *higher-order* kernel, meaning that some of its even moments vanish. In condition (c), the order is P. Higher-order kernels are used in density estimation and nonparametric mean-regression to reduce bias. See the Appendix for further discussion of this use of higher-order kernels. Here, the higher-order kernel is used to make the bias of δ_n have size $o(n^{-1/2})$, which is needed to insure that the asymptotic distribution of $n^{1/2}(\delta_n - \delta)$ is centered at 0. Finally, the rate of convergence of h_n is faster than would be optimal if the aim were to estimate $p(x)$ or $E(Y|x)$ nonparametrically. Under the conditions of Theorem 2.3, the rate of convergence in probability of an estimator of $p(x)$ or $E(Y|x)$ is maximized by setting $h_n \propto n^{-1/(2P+k)}$, which is too slow to satisfy the requirement in condition (d) that $nh_n^{2P} \to 0$ as $n \to \infty$. The relatively fast rate of convergence of h_n required by condition (d), like the higher-order kernel required by condition (c), is needed to prevent the asymptotic distribution of $n^{1/2}(\delta_n - \delta)$ from having a non-zero mean.

Kernel density and mean-regression estimators cannot achieve $O_p(n^{-1/2})$ rates of convergence, so it may seem surprising that δ_n achieves this rate. The fast convergence of δ_n is possible because the sum over i on the right-hand side of (2.46) makes δ_n an average of kernel estimators. Averages of kernel estimators can achieve faster rates of convergence than kernel estimators that are not averaged.

To gain further insight into the reason for the $O_p(n^{-1/2})$ rate of convergence of δ_n as well as intuition about how Theorem 2.3 is proved, define

$$q_n(y_1, x_1, y_2, z_2) = -\left(\frac{1}{h_n}\right)^{k+1} K'\left(\frac{x_1 - x_2}{h_n}\right)(y_1 - y_2).$$

Then

$$(2.48) \qquad \delta_n = \frac{2}{n(n-1)} \sum_{i=1}^{n-1} \sum_{j=i+1}^{n} q_n(Y_i, X_i, Y_j, X_j).$$

The right-hand side of (2.48) is a U statistic, and the theory of U statistics can be used to simplify some of the analysis of δ_n. See Serfling (1980) for a discussion of this theory. The mean of δ_n is

$$E(\delta_n) = -2\int \left(\frac{1}{h_n}\right)^{k+1} K'\left(\frac{x_1 - x_2}{h_n}\right) E(Y|x_1)p(x_1)p(x_2)dx_1dx_2$$

Make the change of variables $x_2 = h_n u + x_1$ to obtain

$$E(\delta_n) = \frac{2}{h_n}\int K'(u)E(Y|x_1)p(x_1)p(x_1 + h_n u)dx_1 du .$$

Integration by parts yields

$$E(\delta_n) = -2\int K(u)E(Y|x)p(x)\frac{\partial p(x + h_n u)}{\partial x}dxdu \tag{2.49}$$

A Taylor series expansion of the integrand on the right-hand side of (2.49) about $h_n = 0$ together with the fact that the moments of K through order P -1 vanish gives

$$E(\delta_n) = \delta + O(h_n^P) . \tag{2.50}$$

To calculate the variance of δ_n, define

$$\tilde{q}_n(y, x) = E[q_n(y, x, Y, X)] . \tag{2.51}$$

Then it follows from the theory of U statistics that

$$\mathrm{Var}(\delta_n) = \frac{4(n-2)}{n(n-1)}\mathrm{Var}[\tilde{q}_n(Y, X)]$$

$$+ O\left\{\frac{1}{n^2}\mathrm{Var}[q_n(Y_1, X_1, Y_2, X_2)]\right\}.$$

Powell, *et al*. (1989) show that $E[q_n(Y_1, X_1, Y_2, X_2)]^2 = O[h_n^{-(k+2)}]$, and (2.50) implies that $E[q_n(Y_1, X_1, Y_2, X_2)] = O(h_n^P)$. Therefore,

$$\mathrm{Var}(\delta_n) = \frac{4}{n}\mathrm{Var}[\tilde{q}_n(Y, X)] + O\left(\frac{1}{n^2 h_n^{k+2}} + h_n^P\right) . \tag{2.52}$$

In addition, Taylor-series arguments similar to those leading to (2.49) show that

$$(2.53) \quad \tilde{q}_n(y,x) = R(y,x) + O(h_n^P).$$

Equation (2.47) is obtained by combining (2.52) and (2.53).

Observe that the remainder terms in (2.50), (2.52), and (2.53) are of size $O(n^{-1/2})$ only if K satisfies condition (c) of Theorem (2.3) and h_n satisfies condition (d). Thus, both of these conditions are needed to make δ_n a $n^{-1/2}$-consistent estimator of δ and to center the asymptotic distribution of $n^{1/2}(\delta_n - \delta)$ at 0.

A consistent estimator of Ω_{AD} can be obtained from (2.47) by replacing δ with δ_n, the population expectation with a sample average, and R with a consistent estimator. To form a consistent estimator of R, combine (2.51) and (2.52) to obtain $R(y,x) = E[q_n(y,x,Y,X)] + O(h_n^P)$. R can be estimated by replacing the population expectation with a sample average in this equation. The result is that Ω_{AD} is estimated consistently by

$$(2.54) \quad \Omega_{AD,n} = \frac{4}{n}\sum_{i=1}^{n} R_n(Y_i, X_i) R_n(Y_i, X_i)' - 4\delta_n\delta_n',$$

where

$$R_n(Y_i, X_i) = -\frac{1}{n-1}\sum_{j \neq i}\left(\frac{1}{h_n}\right)^{k+1} K'\left(\frac{X_i - X_j}{h_n}\right)(Y_i - Y_j).$$

2.7.2 Direct Estimation with Discrete Covariates

Average derivative methods cannot be used to estimate components of β that multiply discrete components of X. This is because derivatives of $E(Y|x)$ with respect to discrete components of X are not identified. This section explains how direct (non-iterative) estimation can be carried out when some components of X are discrete.

To distinguish between continuous and discrete covariates, let X denote the continuously distributed covariates and Z denote the discrete ones. Rewrite (2.1) in the form

$$(2.55) \quad E(Y|X = x, Z = z) = G(x\beta + z\alpha),$$

where α is the vector of coefficients of the discrete covariates. As was discussed in Section 2.4, identification requires that there be at least one continuous covariate. There need not be any discrete covariates, but it is assumed in this section that there is at least one. Let $k_z \geq 1$ denote the number of discrete covariates and components of Z.

The problem of interest in this section is estimating α. The parameter β can be estimated by using density-weighted average derivatives as follows. Let $S_z \equiv \{z^{(i)}: i = 1,..., M\}$ be the points of support of Z. Define $\delta_n^{(i)}$ to be the density-weighted average-derivative estimator of δ (defined in Section 2.7.1) that is obtained by using only observations for which $Z = z^{(i)}$. Let $\delta_{n1}^{(i)}$ be the first component of $\delta_n^{(i)}$. Let w_{ni} $(i = 1,...,M)$ be a set of non-negative (possibly data-dependent) weights that sum to one. The estimator of β is

$$(2.56) \qquad b_n = \frac{\sum_{i=1}^{n} w_{ni}\delta_n^{(i)}}{\sum_{i=1}^{n} w_{ni}\delta_{n1}^{(i)}}.$$

One possible set of weights is $w_{ni} = n_i/n$, where n_i is the number of observations the sample for which $Z = z^{(i)}$. However, the results presented in this section hold with any set of non-negative weights that sum to one.

To see how α can be estimated, assume for the moment that G in (2.55) is known. Let $p(\bullet|z)$ denote the probability density function of $X\beta$ conditional on $Z = z$. Make the following assumption:

Assumption G: There are finite numbers v_0, v_1, c_0, and c_1 such that $v_0 < v_1$, $c_0 < c_1$, and $G(v_0) = c_0$ or c_1 at only finitely many values of v. Moreover, for each $z \in S_z$:

(a) $G(v + z\alpha) < c_0$ if $v < v_0$,

(b) $G(v + z\alpha) > c_1$ if $v > v_1$

(c) $p(\bullet|z)$ is bounded away from 0 on an open interval containing $[v_0, v_1]$.

Parts (a) and (b) of Assumption G impose a form of weak monotonicity on G. G must be smaller than c_0 at sufficiently small values of its argument and larger than c_1 at sufficiently large values. G is unrestricted at intermediate values of its argument. Part (c) insures that $G(v + z\alpha)$ is identified on $v_0 \le v \le v_1$.

To see the implications of Assumption G for estimating α, define

$$J(z) = \int_{v_0}^{v_1} \{c_0 I[G(v+z\alpha) < c_0] + c_1 I[G(v+z\alpha) > c_1]$$

$$+ G(v+z\alpha) I[c_0 \le G(v+z\alpha) \le c_1]\}dv.$$

Define $v_a = \max\{v_0 + z\alpha\colon\ z \in S_z\}$ and $v_b = \min\{v_1 + z\alpha\colon\ z \in S_z\}$. Make the change of variables $v = u - z\alpha$ in the integrals on the right-hand side of $J(z)$. Observe that by Assumption G, $I[G(u) < c_0] = 0$ if $u > v_b$, $I[G(u) > c_1] = 0$ if $u < v_a$, and $I[c_0 \le G(u) \le c_1] = 0$ if $u < v_a$ or $u > v_b$. Therefore,

$$J(z) = c_0 \int_{v_0+z\alpha}^{v_a} I[G(u) < c_0]du + c_0 \int_{v_a}^{v_b} I[G(u) < c_0]du$$

$$+ \int_{v_a}^{v_b} G(u)I[c_0 \le G(u) \le c_1]du + c_1 \int_{v_a}^{v_b} I[G(u) > c_1]du + c_1 \int_{v_b}^{v_1+z\alpha} I[G(u) > c_1]du$$

$$= c_0(v_a - v_0 - z\alpha) + c_0 \int_{v_a}^{v_b} I[G(u) < c_0]du + \int_{v_a}^{v_b} G(u)I[c_0 \le G(u) \le c_1]du$$

$$+ c_1 \int_{v_a}^{v_b} I[G(u) > c_1]du + c_1(v_1 - v_b + z\alpha).$$

It follows that for $i = 2,\dots, M$

$$J[z^{(i)}] - J[z^{(1)}] = (c_1 - c_0)[z^{(i)} - z^{(1)}]\alpha\ . \tag{2.57}$$

Since c_0, c_1, and the support of Z are known, (2.57) constitutes M -1 linear equations in the k_z unknown components of α. These equations can be solved for α if a unique solution exists. To do this, define the $(M - 1)\times 1$ vector ΔJ by

$$\Delta J = \begin{bmatrix} J[z^{(2)}] - J[z^{(1)}] \\ \dots \\ J[z^{(M)}] - J[z^{(1)}] \end{bmatrix}.$$

Also, define the $(M - 1)\times k_z$ matrix W by

$$W = \begin{bmatrix} z^{(2)} - z^{(1)} \\ \dots \\ z^{(M)} - z^{(1)} \end{bmatrix}.$$

Then

$$W'\Delta J = (c_1 - c_0)W'W\alpha .$$

Therefore, if $W'W$ is a nonsingular matrix,

$$\alpha = (c_1 - c_0)(W'W)^{-1}W'\Delta J . \tag{2.58}$$

Equation (2.58) forms the basis of the estimator of α. The estimator is obtained by replacing the unknown $G(v + z\alpha)$ that enters ΔJ with a kernel estimator of the nonparametric mean-regression of Y on Xb_n conditional on $Z = z$. The resulting estimator of $G(v + z\alpha)$ is

$$G_{nz}(v) = \frac{1}{nh_{nz}p_{nz}(v)} \sum_{i=1}^{n} I(Z_i = z)Y_i K\left(\frac{v - V_{ni}}{h_{nz}}\right), \tag{2.59}$$

where h_{nz} is a bandwidth, K is a kernel function, $V_{ni} = X_i b_n$, and

$$p_{nz}(v) = \frac{1}{nh_{nz}} \sum_{i=1}^{n} I(Z_i = z)K\left(\frac{v - V_{ni}}{h_{nz}}\right). \tag{2.60}$$

The estimator of α is then

$$a_n = (c_1 - c_0)(W'W)^{-1}W'\Delta J_n , \tag{2.61}$$

where

$$\Delta J_n = \begin{bmatrix} J_n[z^{(2)}] - J_n[z^{(1)}] \\ \cdots \\ J_n[z^{(M)}] - J_n[z^{(1)}] \end{bmatrix}$$

and

$$J_n(z) = \int_{v_0}^{v_1} \{c_0 I[G_{nz}(v) < c_0] + c_1 I[G_{nz}(v) > c_1]$$

$$+ G_{nz}(v) I[c_0 \le G_{nz}(v) \le c_1)]\}dv.$$

Horowitz and Härdle (1996) give conditions under which a_n in (2.61) is a consistent estimator of α and $n^{1/2}(a_n - \alpha)$ is asymptotically normally distributed with mean 0. The formal statement of this result is given in Theorem 2.4. Define $V = X\beta$, $V_i = X_i\beta$, $v = x\beta$, and $G_z(v) = G(v + z\alpha)$. Let $p(v|z)$ be the probability density of V conditional on $Z = z$, let $p(z)$ be the probability that $Z = z$ ($z \in S_z$), let $p(v,z) = p(v|z)p(z)$, and let $p(v, \tilde{x}|z)$ be the joint density of $(V, \tilde{X})$ conditional on $Z = z$. Finally, define

$$\Gamma_z = -\int_{v_0}^{v_1} G_z'(v) E(\tilde{X}|v,z) I[c_0 \leq G(v + z\alpha) \leq c_1] dv .$$

Theorem 2.4: Let the following conditions hold.

(a) S_z is a finite set. $E(\|\tilde{X}\|^2 | Z = z) < \infty$ and $E(|Y| \|\tilde{X}\|^2 | Z = z) < \infty$ for each $z \in S_z$. $E(|Y|^2 \|\tilde{X}\|^2 | V = v, Z = z)$, $E(|Y|^2 | V = v, Z = z)$, and $p(v,z)$ are bounded uniformly over $v \in [v_0 - \varepsilon, v_1 + \varepsilon]$ for some $\varepsilon > 0$ and all $z \in S_z$. For each $z \in S_z$, $p(v, \tilde{x}|z)$ is everywhere 3 times continuously differentiable with respect to v and the 3rd derivative is bounded uniformly. $\mathrm{Var}(Y|V=v, Z=z) > 0$ for all $z \in S_z$ and almost every v.

(b) $W'W$ is nonsingular.

(c) $E(Y|x,z) = G(x\beta + z\alpha)$. G is r times continuously differentiable for some $r \geq 4$. G and its first r derivatives are bounded on all bounded intervals.

(d) Assumption G holds.

(e) If $k = dim(X) > 1$, there is a $(k - 1)\times 1$ vector-valued function $\omega(y,x,z)$ satisfying $E[\omega(Y, X, Z)] = 0$,

$$n^{1/2}(\tilde{b}_n - \tilde{\beta}) = \frac{1}{n^{1/2}} \sum_{i=1}^{n} \omega(Y_i, X_i, Z_i) + o_p(1),$$

and

$$\frac{1}{n^{1/2}} \sum_{i=1}^{n} \omega(Y_i, X_i, Z_i) \to^d N(0, V_\omega)$$

for some finite matrix V_ω.

(f) K in (2.59)-(2.60) is a bounded, symmetrical, differentiable function that is non-zero only on [-1,1]. K' is Lipschitz continuous. For each integer j between 0 and r ($r \geq 4$)

$$\int_{-1}^{1} v^j K(v)dv = \begin{cases} 1 \text{ if } j = 0 \\ 0 \text{ if } 1 < j < r \\ \text{non-zero if } j = r. \end{cases}$$

(g) As $n \to \infty$, $nh_n^{r+3} \to \infty$ and $nh_n^{2r} \to 0$, where h_n is the bandwidth in (2.59)-(2.60).

Then a_n is a consistent estimator of α. Moreover, $n^{1/2}(a_n - \alpha)$ is asymptotically distributed as $N(0, \Omega_\alpha)$, where Ω_α is the covariance matrix of the $k_z \times 1$ random vector Λ_n whose m'th component is

$$\sum_{j=2}^{m} [(W'W)^{-1}W']_{mj} n^{-1/2} \sum_{i=1}^{n} \{I(Z_i = z^{(j)}) p(V_i, z^{(j)})^{-1}$$

$$\bullet [Y_i - G_{z^{(j)}}(V_i)] I[c_0 \leq G_{z^{(j)}}(V_i) \leq c_1]$$

$$- I(Z_i = z^{(1)}) p(V_i, z^{(1)})^{-1} [Y_i - G_{z^{(1)}}(V_i)] I[c_0 \leq G_{z^{(1)}}(V_i) \leq c_1]$$

$$+ (\Gamma_{z^{(j)}} - \Gamma_{z^{(1)}}) \omega(Y_i, X_i, Z_i)\}.$$

■

Condition (a) makes Z a discrete random variable with finite support and establishes the existence and properties of certain moments. The need for conditions (b) and (d) has already been discussed. Condition (c) makes $E(Y|x, z)$ a single-index model. Condition (e) is satisfied by the estimator of β defined in (2.56) but does not require the use of this estimator. Conditions (f) and (g) require K to be a higher-order kernel with undersmoothing. As in density-weighted average-derivative estimation, conditions (f) and (g) are used to insure that the asymptotic distribution of $n^{1/2}(a_n - \alpha)$ is centered at 0.

The covariance matrix Ω_α can be estimated consistently by replacing unknown quantities with consistent estimators. Γ_z is estimated consistently by

$$\Gamma_{nz} = -\frac{1}{n}\sum_{i=1}^{n}\tilde{X}_i 1(Z_i = z) I(v_0 \leq V_{ni} \leq v_1) I[c_0 \leq G_{nz}(V_{ni}) \leq c_1)$$

$$\bullet G'_{nz}(V_{ni}) / p_{nz}(V_{ni}),$$

where $G'_{nz}(v) = dG_{nz}(v)/dv$. Define $\lambda(y,v,z)$ to be the $(M - 1)\times 1$ vector whose $(j - 1)$ component $(j = 2,..., M)$ is

$$\lambda_j(y,v,z) = I(z = z^{(j)})\frac{y - G_{nz^{(j)}}(v)}{p_{nz^{(j)}}(v)} I[c_0 \leq G_{nz^{(j)}}(v) \leq c_1]$$

$$- I(z = z^{(1)})\frac{y - G_{nz^{(1)}}(v)}{p_{nz^{(1)}}(v)} I[c_0 \leq G_{nz^{(1)}}(v) \leq c_1].$$

Let ω_n be a consistent estimator of ω. Then Ω_α is estimated consistently by the sample covariance of the $k_z\times 1$ vector whose m'th component $(m = 1,..., k_z)$ is

$$\sum_{j=2}^{m}[(W'W)^{-1}W']_{mj}[\lambda_j(Y_i, V_{ni}, Z_i) + (\Gamma_{nz^{(j)}} - \Gamma_{nz^{(1)}})\omega_n(Y_i, X_i, Z_i)].$$

Horowitz and Härdle (1996) show how to estimate ω when the estimator of β is (2.56). To state their result, let $p_{ni}(x)$ be a kernel estimator of the probability density of X conditional on $Z = z^{(i)}$. That is,

$$p_{ni}(x) = \frac{1}{n_i s_n}\sum_{j=1}^{n} 1(Z_j = z^{(i)}) K^*\left(\frac{x - X_j}{s_n}\right),$$

where K^* is a kernel function of a k-dimensional argument, n_i is the number of observations for which $Z = z^{(i)}$, and s_n is a bandwidth. Let $x^{(1)}$ be the first component of x. Then the estimator of ω is

$$\omega_n(y,x,z^{(i)}) = -2\frac{n_i}{n\delta_{n1}^{(i)}}[y - G(xb_n + z^{(i)}a_n)]\left[\frac{\partial p_{ni}(x)}{\partial \tilde{x}} - \tilde{b}_n \frac{\partial p_{ni}(x)}{\partial x^{(1)}}\right].$$

2.7.3 One-Step Asymptotically Efficient Estimators

In parametric estimation, an asymptotically efficient estimator can be obtained by taking one Newton step from any $n^{1/2}$-consistent estimator toward the maximum likelihood estimator. This procedure is called *one-step asymptotically efficient estimation*. The resulting estimator is called a *one-step asymptotically efficient estimator*. This section shows that the same idea applies to estimation of β in a semiparametric single-index model. Specifically, let $S_n(\bullet)$ be the objective function of the semiparametric WNLS estimator (2.30) with $W(x) = 1/s_n^2(x)$. Then an asymptotically efficient estimator of β can be obtained by taking one Newton step from any $n^{1/2}$-consistent estimator toward the minimum of S_n. In the case of a single-index binary-response model, the step may be taken toward the maximum of the semiparametric log-likelihood function (2.39).

One-step asymptotically efficient estimation is especially useful in semiparametric single-index models because the direct estimators described in Sections 2.7.1 and 2.7.2 can be computed very rapidly. Therefore, one-step estimators can be obtained with much less computation than is needed to minimize S_n or maximize the semiparametric log-likelihood function.

Consider one-step asymptotically efficient estimation based on S_n. Let X denote the entire vector of covariates, continuous and discrete. Let β denote the entire vector of coefficients of X in (2.1). Let $\tilde{b}_n^*$ be any $n^{1/2}$-consistent estimator of $\tilde{\beta}$. It is convenient in applications but not essential to the arguments made here to let $\tilde{b}_n^*$ be a direct estimator. Let S_n be the semiparametric WNLS objective function (2.30) with $W(x) = 1/s_n^2(x)$. The one-step asymptotically efficient estimator of $\tilde{\beta}$ is

$$(2.62)\qquad \tilde{b}_n = \tilde{b}_n^* - \left[\frac{\partial^2 S_n(\tilde{b}_n^*)}{\partial\tilde{b}\,\partial\tilde{b}'}\right]^{-1} \frac{\partial S_n(\tilde{b}_n^*)}{\partial\tilde{b}}$$

To see why $\tilde{b}_n$ is asymptotically efficient, write (2.62) in the form

$$(2.63)\qquad n^{1/2}(\tilde{b}_n - \tilde{\beta}) = n^{1/2}(\tilde{b}_n^* - \tilde{\beta}) - \left[\frac{\partial^2 S_n(\tilde{b}_n^*)}{\partial\tilde{b}\,\partial\tilde{b}'}\right]^{-1} n^{1/2}\frac{\partial S_n(\tilde{b}_n^*)}{\partial\tilde{b}}$$

Observe that just as in the arguments leading to (2.36),

$$(2.64)\qquad \frac{\partial^2 S_n(\tilde{b}_n^*)}{\partial \tilde{b}\, \partial \tilde{b}'} = C + o_p(1)\ .$$

Moreover, a Taylor series expansion gives

$$\frac{\partial S_n(\tilde{b}_n^*)}{\partial \tilde{b}} = \frac{\partial S_n(\tilde{\beta})}{\partial \tilde{b}} + \frac{\partial^2 S_n(\bar{b}_n)}{\partial \tilde{b}\, \partial \tilde{b}'}(\tilde{b}_n^* - \tilde{\beta})\,,$$

where $\bar{b}_n$ is between $\tilde{b}_n^*$ and $\tilde{\beta}$. The second-derivative term in this equation converges in probability to C, so

$$(2.65)\qquad n^{1/2}\frac{\partial S_n(\tilde{b}_n^*)}{\partial \tilde{b}} = n^{1/2}\frac{\partial S_n(\tilde{\beta})}{\partial \tilde{b}} + Cn^{1/2}(\tilde{b}_n^* - \tilde{\beta}) + o_p(1)\,.$$

Substitution of (2.64) and (2.65) into (2.63) yields

$$n^{1/2}(\tilde{b}_n - \tilde{\beta}) = -C^{-1}n^{1/2}\frac{\partial S_n(\tilde{\beta})}{\partial \tilde{b}} + o_p(1)\,.$$

As in (2.35)

$$n^{1/2}\frac{\partial S_n(\tilde{\beta})}{\partial \tilde{b}} \to^d N(0, D)\,.$$

Therefore, $n^{1/2}(\tilde{b}_n - \tilde{\beta}) \to^d N(0, C^{-1}DC^{-1})$. Since $C^{-1}DC^{-1} = \Omega_{SI}$ when $W(x) = 1/s_n^2(x)$,

$$n^{1/2}(\tilde{b}_n - \tilde{\beta}) \to^d N(0, \Omega_{SI})\,.$$

This establishes the asymptotic efficiency of the one-step semiparametric WNLS estimator. The same arguments apply to the one-step semiparametric maximum-likelihood estimator after replacing S_n with the semiparametric log-likelihood function.

2.8 Bandwidth Selection

Implementation of any of the semiparametric estimators for single-index models that are discussed in this chapter requires choosing the numerical values of one or more bandwidth parameters and, possibly, of other tuning parameters. Except for the case of density-weighted average-derivative estimation, there has been little research on how this should be done.

As can be seen from the results in Sections 2.6-2.7, in semiparametric single-index models, the asymptotic distribution of $n^{1/2}(b_n - \beta)$ does not depend on the bandwidth h_n. Therefore, bandwidth selection must be based on a higher-order approximation to the distribution of $n^{1/2}(b_n - \beta)$. Härdle and Tsybakov (1993) used such an approximation to obtain a formula for the bandwidth that minimizes the asymptotic approximation to $E\|\delta_n - \delta\|^2$, where δ and δ_n, respectively, are as in (2.43) and (2.46), and $\|\bullet\|$ is the Euclidean norm. This is called the *asymptotically optimal* bandwidth. Powell and Stoker (1996) obtained the bandwidth that minimizes the asymptotic mean-square error of a single component of $\delta_n - \delta$.

Two aspects of the results of Härdle and Tsybakov (1993) and Powell and Stoker (1996) are especially noteworthy. First, the asymptotically optimal bandwidth has the form

$$(2.66) \qquad h_{n,opt} = h_0 n^{-2/(2P+k+2)},$$

where P and k are defined as in Theorem 2.3, h_0 is a constant. Second, Powell and Stoker (1996) provide a method for estimating h_0 in an application. To state this method, let h_{n1} be an initial bandwidth estimate that satisfies $h_{n1} \to 0$ and $nh_{n1}{}^{c} \to \infty$ as $n \to \infty$, where $c = \max(\eta + 2k + 4, 2P + k + 2)$ for some $\eta > 0$. Define

$$q_n(y_1, x_1, y_2, z_2) = -\left(\frac{1}{h_{n1}}\right)^{k+1} K'\left(\frac{x_1 - x_2}{h_{n1}}\right)(y_1 - y_2)$$

and

$$\hat{Q} = \frac{2h_{n1}^{k+2}}{n(n-1)} \sum_{i<j} q_n(Y_i, X_i, Y_j, X_j)^2 .$$

Let $\delta_n(h)$ denote the density-weighted average-derivative estimator of δ based on bandwidth h. Let $\tau \neq 1$ be a positive number. Define

$$\hat{S} = \frac{\delta_n(\tau h_{n1}) - \delta_n(h_{n1})}{(\tau h_{n1})^P - h_{n1}^P} .$$

The estimator of h_0 is

(2.67) $$\hat{h}_0 = \left[\frac{(k+2)\hat{Q}}{P\hat{S}^2} \right]^{1/(2P+k+2)} .$$

Another possible approach to bandwidth selection is based on resampling the data. Suppose that the asymptotically optimal bandwidth has the form

$$h_{n,opt} = h_0 n^{-\gamma}$$

for some known γ. For example, in density-weighted average-derivative estimation, $\gamma = 2P + k + 2$. Let $m < n$ be a positive integer. Let $\{Y_i^*, X_i^*: i = 1,..., m\}$ be a sample of size m that is obtained by sampling the estimation data randomly without replacement. Then $\{Y_i^*, X_i^*\}$ is a random sample from the population distribution of (Y, X). Repeat this resampling process J times. Let $b_{mj}(h)$ $(j = 1,..., J)$ be the estimate of β that is obtained from the j'th sample using bandwidth $h = \tau m^{-\gamma}$, where τ is a constant. Let b_n be the estimate of β that is obtained from the full sample by using a preliminary bandwidth estimate that satisfies the requirements needed to make b_n a $n^{1/2}$-consistent estimator of β. Let τ_m be the solution to the problem

$$\underset{\tau}{\text{minimize:}} \ \frac{1}{J} \sum_{j=1}^{J} \left[b_{mj}(h) - b_n \right]^2 .$$

Then τ_m estimates h_0, and $h_{n,opt}$ is estimated by

$$\hat{h}_{n,opt} = \tau_m n^{-\gamma} .$$

Horowitz and Härdle (1996) used Monte Carlo methods to obtain rules of thumb for selecting the tuning parameters required for the estimator of α described in Section 2.7.2. Horowitz and Härdle (1996) obtained good numerical results in Monte Carlo experiments by setting $h_{nz} = s_{vz} n_z^{-1/7.5}$, where s_{vz} is the sample standard deviation of Xb_n conditional on $Z = z \in S_z$ and n_z is the number of observations with $Z = z$. In these experiments, the values of the other tuning parameters were

$$v_1 = \min_{z \in S_z} \max_{1 \le i \le n} \{X_i b_n - h_{nz} : Z_i = z\} ,$$

$$v_0 = \max_{z \in S_z} \min_{1 \le i \le n} \{X_i b_n + h_{nz} : Z_i = z\} ,$$

$$c_0 = \max_{z \in S_z} \max_{X_i b_n \leq v_0} G_{nz}^*(X_i b_n),$$

and

$$c_1 = \min_{z \in S_z} \min_{X_i b_n \geq v_0} G_{nz}^*(X_i b_n).$$

In the formulae for c_0 and c_1, G_{nz}^* is the kernel estimator of G_z that is obtained using a second-order kernel instead of the higher-order kernel used to estimate $\alpha_{\cdot}$. Horowitz and Härdle (1996) found that using a second-order kernel produced estimates of c_0 and c_1 that were more stable than those obtained with a higher-order kernel.

2.9 An Empirical Example

This section presents an empirical example that illustrates the usefulness of semiparametric single-index models. The example is taken from Horowitz and Härdle (1996) and consists of estimating a model of product innovation by German manufacturers of investment goods. The data, assembled in 1989 by the IFO Institute of Munich, consist of observations on 1100 manufacturers. The dependent variable is $Y = 1$ of a manufacturer realized an innovation during 1989 in a specific product category and 0 otherwise. The independent variables are the number of employees in the product category (EMPLP), the number of employees in the entire firm (EMPLF), an indicator of the firm's production capacity utilization (CAP), and a discrete variable DEM, which is 1 if a firm expected increasing demand in the product category and 0 otherwise. The first three independent variables are standardized so that they have units of standard deviations from their means. Scale normalization was achieved by setting $\beta_{EMPLP} = 1$.

Table 2.3 shows the parameter estimates obtained using a binary probit model and the direct semiparametric methods of Sections 2.7.1 and 2.7.2.

Table 2.3: Estimated Coefficients (Standard Errors) for Model of Product Innovation

EMPLP	EMPLF	CAP	DEM
	Semiparametric Model		
1	0.032	0.346	1.732
	(0.023)	(0.078)	(0.509)
	Probit Model		
1	0.516	0.520	1.895
	(0.024)	(0.163)	(0.387)

Source: Horowitz and Härdle (1996). The coefficient of EMPLP is 1 by scale normalization.

Figure 2.1 shows a kernel estimate of $G'(v)$. There are two important differences between the semiparametric and probit estimates. First, the semiparametric estimate of β_{EMPLF} is small and statistically nonsignificant, whereas the probit estimate is significant at the 0.05 level and similar in size to β_{CAP}. Second, in the binary probit model, G is a cumulative normal distribution function, so G' is a normal density function. Figure 2.1 reveals, however, that G' is bimodal. This bimodality suggests that the data may be a mixture of two populations. An obvious next step in the analysis of the data would be to search for variables that characterize these populations. Standard diagnostic techniques for binary probit models would provide no indication that G' is bimodal. Thus, the semiparametric estimate has revealed an important feature of the data that could not easily be found using standard parametric methods.

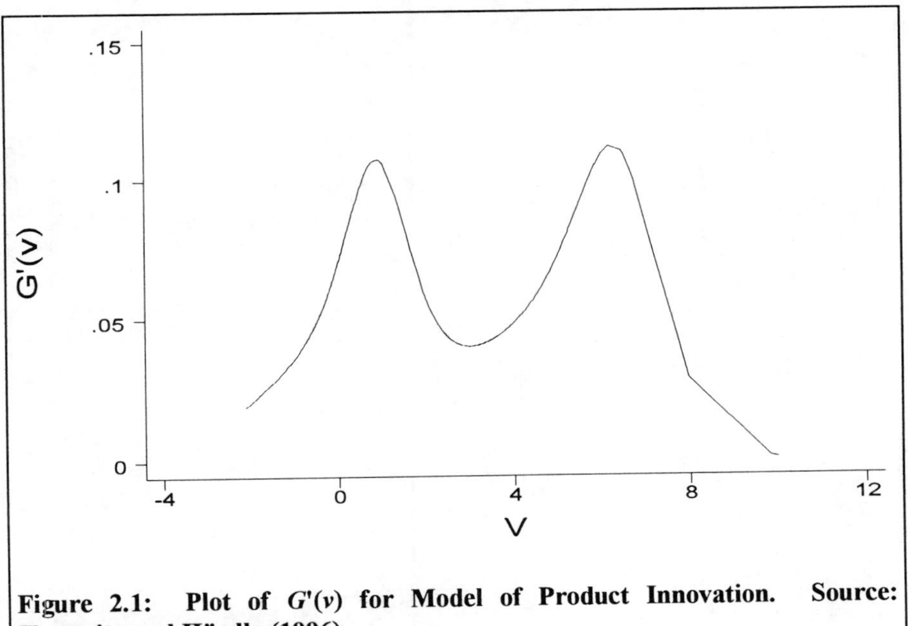

Figure 2.1: Plot of $G'(v)$ for Model of Product Innovation. Source: Horowitz and Härdle (1996).

Chapter 3
Binary Response Models

This chapter is concerned with estimating the binary-response model

(3.1a) $$Y = \begin{cases} 1 & \text{if } Y^* > 0 \\ 0 & \text{otherwise} \end{cases}$$

where

(3.1b) $$Y^* = X\beta + U ,$$

Y is the observed dependent variable, X is a $1 \times k$ vector of observed explanatory variables, β is a $k \times 1$ vector of constant parameters, Y^* is an unobserved, latent dependent variable, and U is an unobserved random variable. The inferential problem is to use observations of (Y, X) to estimate β and, to the extent possible, the probability that $Y = 1$ conditional on X.

If the distribution of U belongs to a known, finite-dimensional parametric family, β and any parameters of the distribution of U can be estimated by maximum likelihood. As was demonstrated in Section 2.9, however, the results of maximum likelihood estimation can be highly misleading if the distribution of U is misspecified. If the distribution of U does not belong to a known finite-dimensional family but is known to be independent of X or to depend on X only through the index $X\beta$, then (3.1) is a semiparametric single-index model. The parameters β can be estimated using the methods of Chapter 2. As was discussed in Chapter 2, semiparametric single-index models have wide applicability. They do not, however, permit a form of heteroskedasticity of U called *random coefficients* that is often important in applications. This chapter presents a semiparametric binary-response model that accommodates many different kinds of heteroskedasticity, including random coefficients.

3.1 Random-Coefficients Models

To understand what a random-coefficients model is and why such models can be important in applications, suppose that (3.1) is a model of an individual's choice between two alternatives, say traveling to work by automobile and traveling to work by bus. Automobile is chosen if $Y = 1$, and bus is chosen otherwise. Suppose that X represents differences between attributes of automobile and bus travel that are relevant to choice. For example, one component of X might be the difference between the costs of travel to work by automobile and bus, another component might be the difference between travel times by automobile and bus, etc. The parameter β then characterizes an individual's preferences or tastes regarding the attributes in X. It is entirely possible that different individuals have different tastes for reasons that cannot be explained by X or other variables that are observed by an analyst. Then from the point of view of the analyst the parameters β of model (3.1) vary randomly among individuals, thereby making (3.1) a random-coefficients model. Hausman and Wise (1978), Fischer and Nagin (1981), and Horowitz (1993a) present empirical models of travel choices in which the coefficients are random.

The formal structure of a random-coefficients, binary-response model is

$$(3.2a)\qquad Y = \begin{cases} 1 & \text{if } Y^* \geq 0 \\ 0 & \text{otherwise} \end{cases}$$

where

$$(3.2b)\qquad Y^* = X(\beta + \nu) + V,$$

$$(3.2c)\qquad = X\beta + (X\nu + V)$$

$$(3.2d)\qquad = X\beta + U,$$

and $U = X\nu + V$. In (3.2b), the coefficient of X is $\beta + \nu$, where β is a constant vector and ν is a random vector. The constant β is the mean or median of the distribution of the coefficient of X. The random vector ν is unobserved and accounts for deviations of the coefficient of X from its population mean or median. Equations (3.2a) and (3.2d) show that the random-coefficients model has the form of the general binary-response model (3.1) but with a heteroskedastic random component U.

The problem of estimating β without assuming that the distributions of ν and V belong to known parametric families will now be addressed. The methods that will be developed permit a wide variety of different forms of heteroskedasticity of U, not just random coefficients.

3.2 Identification

The first task is to find conditions under which β is identified. That is, it is necessary to determine when β is uniquely determined by the population distribution of (Y, X).

To begin, observe that $P(Y = 1|x)$ is unchanged if both sides of (3.1b) are multiplied by a positive constant. Therefore, β can only be identified up to scale, and a scale normalization is needed. In this chapter, as in Chapter 2, scale normalization will be achieved by setting the first component of β equal to 1. For reasons that will be explained below, it will be assumed that the first component of X is a continuously distributed random variable.

It can also be seen from (3.1) that $P(Y = 1|x)$ is unchanged if a constant (positive or negative) is added to both sides of (3.1b). Therefore, a location normalization is needed. The most familiar location normalization sets the mean of U conditional on $X = x$ equal to zero: $E(U|x) = 0$. Under this location normalization, β in a random-coefficients model is the mean of the coefficient of X. If Y^* were observable, so that (3.1b) were a linear model, then the assumptions that $E(U|x) = 0$ and that there is no exact linear relation among the components of X would be sufficient to identify β. It turns out, however, that in a binary-response model, where Y^* is not observable, the location normalization $E(U|x) = 0$ is not sufficient to identify β even if the components of X are not linearly related. This non-identification result was first obtained by Manski (1985, 1988). The following example, which is given in Horowitz (1993b), illustrates the problem.

Example 3.1: A Binary-Response Model with Mean Independence

Suppose that $P(Y = 1|x)$ is given by a the binary logit model

$$P(Y = 1|x) = \frac{1}{1+\exp(-x\beta)}.$$

This model can be obtained from (3.1) by assuming that U has the standard logistic distribution. The CDF of the standard logistic distribution is

$$F_L(u) = \frac{1}{1+\exp(-u)}.$$

It is easy to show that $E(U) = 0$ if U has this CDF.

Now let $b \neq \beta$ be any parameter value that satisfies the scale normalization. It is shown below that for each x in the support of X, there is a random variable V_x with CDF $F_{V|x}$ such that $E(V_x|x) = 0$ and

$$(3.3) \qquad F_{V|x}(xb) = \frac{1}{1+\exp(-x\beta)}.$$

Therefore $P(Y = 1|x)$ can be obtained from a binary logit model with parameter β and from the model consisting of (3.1a) and

$$Y^* = Xb + V_x,$$

where $E(V_x|x) = 0$. This implies that β is not identified because identical choice probabilities are obtained with parameter value β and parameter value $b \neq \beta$.

To see how the random variable V_x can be constructed, let x be given. Consider the random variable W whose CDF conditional on $X = x$ is

$$F_W(w|x) = \frac{1}{1+\exp[-w + x(b-\beta)]}.$$

Then

$$F_W(xb|x) = \frac{1}{1+\exp(-x\beta)}$$

and $E(W|x) = x(b - \beta) \equiv \delta_x$. Suppose that $\delta_x > 0$. Then construct $F_{V|x}$ from the distribution of W by taking part of the probability mass of W that is to the left of xb and moving it enough further to the left to make the resulting distribution have mean 0 conditional on $X = x$. If $\delta_x < 0$, construct $F_{V|x}$ by moving probability mass rightward from the part of the distribution of W that is to the right of xb. Since no probability mass crosses the point $W = xb$ in these movements, the resulting distribution satisfies (3.3) with $E(V_x|x) = 0$ as required. ■

Example 3.1 shows that the mean-independence assumption $E(U|x) = 0$ does not restrict the distribution of U sufficiently to identify β. Identification is possible if mean independence is replaced with the stronger assumption that U is statistically independent of X. However, statistical independence of U and X precludes heteroskedasticity and, therefore, is too strong an assumption for the models of interest in this chapter. A location normalization condition that permits heteroskedasticity and identifies β is *median independence* or median$(U|x) = 0$. More generally, it suffices to assume that any quantile of the distribution of U is independent of X, but only median independence will be discussed here. The following theorem gives conditions that are sufficient for identification of β.

Theorem 3.1 (Identification under median independence): Let median($U|x$) = 0 for all x in the support of X. Let the first component if β equal 1. Then β is identified if

(a) The support of the distribution of X is not contained in any proper linear subspace of $\Re^k$.

(b) For almost every $\tilde{x} = (x_2, \ldots, x_k)$ the distribution of X_1 conditional on $\tilde{X} = \tilde{x}$ has an everywhere positive density. ■

Theorem 3.1 is a slightly modified version of an identification result that was originally obtained by Manski (1985). Manski did not impose the scale normalization $\beta_1 = 1$. Instead, he showed that β is identified up to scale under conditions (a) and (b) of Theorem 3.1.

Condition (a) implies that there is no exact linear relation among the components of X. To understand the role of condition (b) and why median($U|x$) = 0 permits identification, let $\tilde{\beta}$ and $\tilde{b}$, respectively, denote the vectors consisting of all components of β and b but the first. Observe that $P(Y = 1|x) = P(U > -x\beta|x) = 1 - P(U \leq -x\beta|x)$. Therefore,

$$P(Y = 1|x) \geq 0.5 \text{ if } x\beta \geq 0$$

$$< 0.5 \text{ if } x\beta < 0.$$

Let $b \neq \beta$ be a parameter value that satisfies the scale normalization. Let $S_1(b)$ and $S_2(b)$ be the following sets:

$$S_1(b) = \{x: x\beta < 0 \leq xb\} \tag{3.4}$$

$$S_2(b) = \{x: xb < 0 \leq x\beta\}. \tag{3.5}$$

If $P[S_1(b) > 0]$, then b is observationally distinguishable from β because there is a subset of the support of X that occurs with nonzero probability and on which $P(Y = 1|x) < 0.5$ with parameter value β but $P(Y = 1|x) \geq 0.5$ with parameter value b. Similarly, if $P[S_2(b) > 0]$, then b is observationally distinguishable from β because there is a subset of the support of X that occurs with nonzero probability and on which $P(Y = 1|x) \geq 0.5$ with parameter value β but $P(Y = 1|x) < 0.5$ with parameter value b. Therefore, β is identified if

$$P[S_1(b) \cup S_2(b)] > 0 \tag{3.6}$$

for all $b \neq \beta$. Indeed, Manski (1988) has shown that (3.6) is necessary as well as sufficient for identification of β relative to b. The necessity of (3.6) will be used in Sections 3.2.1 and 3.2.2, which deal with identification when the support of X is bounded or discrete.

Because the first components of b and β equal 1, for any $b \neq \beta$,

$$S_1(b) = \{x: -\tilde{x}\tilde{b} \leq x_1 < -\tilde{x}\tilde{\beta})$$

and

$$S_2(b) = \{x: -\tilde{x}\tilde{\beta} \leq x_1 < -\tilde{x}\tilde{b}) .$$

If the distribution of X_1 has everywhere positive density conditional on $\tilde{X} = \tilde{x}$, then $S_1(b)$ has positive probability whenever $-x\tilde{b} < -x\tilde{\beta}$, and $S_2(b)$ has positive probability whenever $-\tilde{x}\tilde{\beta} < -\tilde{x}\tilde{b}$. Therefore, β is identified if

$$\text{(3.7)} \qquad P(\tilde{X}\tilde{b} = \tilde{X}\tilde{\beta}) < 1$$

Inequality (3.7) holds whenever condition (a) of Theorem 3.1 holds, thereby establishing identification if β under the assumptions of Theorem 3.1.

Condition (b) of Theorem 3.1 implies that X has at least one continuously distributed component, X_1, and that this component has unbounded support. It is worth considering whether identification is possible if these conditions fail to hold. In the next section, it is shown that unbounded support of X_1 is not necessary for identification of β. Continuity is necessary except in special cases.

3.2.1 Identification Analysis when X Has Bounded Support

In this section it is assumed that X has bounded support but that the distribution of X_1 conditional on $\tilde{X}$ has a density. Condition (b) of Theorem 3.1 is violated because the density of X_1 conditional on $\tilde{X}$ is not everywhere positive.

Let S_x denote the bounded support of X. It can be seen from (3.4)-(3.6) that β is not identified if there is a $\delta > 0$ such that $|x\beta| \geq \delta$ for all $x \in S_x$. If $|x\beta| \geq \delta$ for all $x \in S_x$, then xb has the same sign as $x\beta$ for all $x \in S_x$ and any b that is sufficiently close to β. Therefore, $S_1(b)$ and $S_2(b)$ are empty sets for all such b's, and (3.6) does not hold for them. Since (3.6) is necessary for identification of β relative to b, β is not identified if $|x\beta| \geq \delta$ for all $x \in S_x$. This result was first obtained by Manski (1985).

Identification is possible, however, if the support of $X\beta$ conditional on $\tilde{X}$ includes an interval containing $X\beta = 0$ for sufficiently many values of $\tilde{X}$. The following corollary to Theorem 3.1 gives conditions sufficient for identification of β when X has bounded support.

Corollary 3.1.1 (Identification of β when X has bounded support): Let X have bounded support. Let median($U|x$) = 0 for all x in the support of X. Let the first component if β equal 1. Then β is identified if for some $\delta > 0$, there are an interval $I_\delta = [-\delta, \delta]$ and a set $N_\delta \subset \Re^{k-1}$ such that

(a) N_δ is not contained in any proper linear subspace of $\Re^{k-1}$.

(b) $P(\tilde{X} \in N_\delta) > 0$

(c) For almost every $\tilde{x} \in N_\delta$, the distribution of $X\beta$ conditional on $\tilde{X} = \tilde{x}$ has a probability density that is everywhere positive on I_δ. ■

Conditions (a)-(c) of Corollary 3.1.1 imply that for each $\tilde{x}$ in a sufficiently rich set, there is an interval of the real line with the property that $P(Y = 1|X_1 = x_1, \tilde{X} = \tilde{x}) > 0.5$ for some values of x_1 in this interval and $P(Y = 1|X_1 = x_1, \tilde{X} = \tilde{x}) \leq 0.5$ for others. If $P(Y = 1|x)$ is a continuous function of the components of x corresponding to continuous components of X, this implication can be tested, at least in principle, by estimating the nonparametric mean regression of Y on X.

To understand why conditions (a)-(c) of Corollary 3.1.1 are sufficient to identify β, observe that the sets $S_1(b)$ and $S_2(b)$ defined in (3.4) and (3.5) can be written in the form

$$S_1(b) = \{x: -\tilde{x}(\tilde{b} - \tilde{\beta}) \leq x\beta < 0\}$$

and

$$S_2(b) = \{x: 0 \leq x\beta < -\tilde{x}(\tilde{b} - \tilde{\beta})\}.$$

Condition (c) of Corollary (3.1.1) implies that at least one of these sets has nonzero probability if $\tilde{x} \in N_\delta$ and $\tilde{x}(\tilde{b} - \tilde{\beta}) \neq 0$. Conditions (a) and (b) imply that $P[\tilde{X}(\tilde{b} - \tilde{\beta}) \neq 0] > 0$ if $b \neq \beta$. Therefore, condition (3.6) holds, and β is identified.

If $|x\beta| > \delta$ for some $\delta > 0$ and all $x \in S_x$, then β is not identified but it may be possible to bound individual components of β. Manski (1988) has given the following conditions under which the signs of components of β are identified.

Theorem 3.2 (Manski 1988): Let X^* be the vector of all components of X except the m'th ($m \neq 1$). Let S^* denote the support of X^*. For any $b \in \Re^k$, Define

$$S_b = \{x^* \in S^*: P(Xb < 0 | X^* = x^*) > 0 \text{ or } P(Xb > 0 | X^* = x^*) > 0\}.$$

If $\beta_m \neq 0$, then the sign of β_m is identified if $P(X^* \in S_\beta) > 0$. If $\beta_m = 0$, then β_m is identified if $P(X^* \in S_b) > 0$ for all $b \in \Re^k$ such that $b_m \neq 0$. ■

To obtain further results on bounding β, suppose that $P(Y = 1|x)$ is a continuous function of the components of x that correspond to continuous components of X. Under this continuity assumption, $P(Y = 1|x)$ is identified by the nonparametric mean regression of Y on X. Therefore, bounds (possibly infinite) on β_m ($m \neq 1$) can be obtained by solving the problems

$$\begin{aligned} &\text{maximize (minimize): } b_m \\ &\text{subject to: } xb \leq 0 \text{ if } P(Y = 1 | x) < 0.5 \\ &\qquad\qquad\quad xb \geq 0 \text{ if } P(Y = 1 | x) \geq 0.5 \\ &\text{for all } x \in S_x \end{aligned} \tag{3.8}$$

Problem (3.8) finds the largest and smallest values of b_m that allow the signs of xb and $x\beta$ to be the same for every $x \in S_x$. The sign of $x\beta$ is identified because it is determined by whether the identified quantity $P(Y = 1|x)$ does or does not exceed 0.5.

The linear programming problem (3.8) has infinitely many constraints, so solving it in practice can be difficult. A version of (3.8) that can be solved with standard linear programming techniques can be obtained by considering only the constraints for which x takes values in a finite subset of S_x, possibly the values contained in the available data. The resulting bounds are looser than the ones obtained by solving the infinitely constrained problem but are arbitrarily close to the bounds obtained from the infinitely constrained (3.8) if the chosen subset is sufficiently dense.

Example 3.3 in Section 3.2.2 illustrates the use of (3.8).

3.2.2 Identification when X Is Discrete

In this section, it is assumed that the support of X, S_x, consists of finitely many discrete points. Suppose, first, that $|x\beta| > 0$ for all $x \in S_x$. Then arguments identical to those made at the beginning of Section 3.2.1 show that there are points b for which (3.6) does not hold. Therefore, β is not identified.

The situation is more complicated if $x\beta = 0$ for one or more points $x \in S_x$. To analyze this case, define $S_0 = \{x\text{: } x\beta = 0, x \in S_x\}$. Let N_0 be the number of points in S_0. Because S_x is finite, $|x\beta| \geq \delta$ for all $x \in S_x$ - S_0 and some $\delta > 0$. Choose M so that $||x|| \leq M$ for all $x \in S_x$, and choose ε such that $0 < \varepsilon < \delta/M$. Then $x\beta$ and xb have the same sign if $x \in S_x - S_0$ and $||b - \beta|| \leq \varepsilon$. Therefore, if $||b - \beta|| \leq \varepsilon$, $S_1(b)$ is empty and

$$\text{(3.9)} \qquad S_2(b) = \{x\text{: } xb < 0, x \in S_0\}\,.$$

Because (3.6) is necessary for identification of β relative to b, β is not identified if there is any b such that $||b - \beta|| \leq \varepsilon$ and $S_2(b)$ in (3.9) is empty.

It is not difficult to show that such b's always exist if $N_0 \leq k - 1$, where $k = \dim(X)$. Let $\hat{x}$ be the $N_0 \times k$ matrix that is obtained by stacking the elements of S_0. Let e be a $N_0 \times 1$ vector of ones. Let $c > 0$ be a scalar. Consider the equation

$$\text{(3.10)} \qquad \hat{x}b^* = c$$

Because $\hat{x}\beta = 0$, (3.10) is equivalent to

$$\text{(3.11)} \qquad \hat{x}(b^* - \beta) = c$$

Scale normalization equates the first components of b^* and β, so (3.11) is N_0 linear equations with k -1 unknown quantities. Let $(b^* - \beta)$ be a solution. If $||b^* - \beta|| \leq \varepsilon$, set $b = b^*$. Then $\hat{x}b > 0$, $S_2(b)$ is empty, and β is not identified relative to this b. If $||b^* - \beta|| = \lambda > \varepsilon$, set $b = (1, \tilde{b}')'$, where

$$\tilde{b} = \tilde{\beta} + \frac{\varepsilon}{\lambda}(\tilde{b}^* - \tilde{\beta})$$

and $\tilde{b}^*$ denotes components 2 through k of b^*. Then $||b - \beta|| = \varepsilon$ and

$$\hat{x}b = \hat{x}\beta + \frac{\varepsilon}{\lambda}\hat{x}(b^* - \beta)$$

$$= \frac{\varepsilon}{\lambda}c$$

$$> 0.$$

It follows that $S_2(b)$ is empty, and β is not identified relative to this b. Therefore, if $N_0 \leq k - 1$, there are always points b for which $S_2(b)$ is empty. It follows that β is not identified if $N_0 \leq k - 1$.

If $N_0 \geq k$, β may or may not be identified, depending on the configuration of points in S_0. This is illustrated by the following example.

Example 3.2: Identification of β when X is Discrete

Let $k = 2$. Let $S_0 = \{(1,-1), (-1,1)\}$ and $\beta = (1, 1)'$. Then

$$\hat{x}b = \begin{bmatrix} 1 & -1 \\ -1 & 1 \end{bmatrix} \begin{bmatrix} 1 \\ \tilde{b} \end{bmatrix}$$

$$= \begin{bmatrix} 1 - \tilde{b} \\ -1 + \tilde{b} \end{bmatrix}$$

for any b. Therefore, $\hat{x}b$ always has a negative element when $b \neq \beta$, and β is identified.

Now let $S_0 = \{(1,1), (2,2)\}$ and $\beta = (1, -1)'$. Then

$$\hat{x}b = \begin{bmatrix} 1 & 1 \\ 2 & 2 \end{bmatrix} \begin{bmatrix} 1 \\ \tilde{b} \end{bmatrix}$$

$$= \begin{bmatrix} 1 + \tilde{b} \\ 2 + 2\tilde{b} \end{bmatrix}$$

Given any $\varepsilon > 0$, let $b = (1, -1 + \varepsilon/2)'$. Then

$$\hat{x}b = \begin{bmatrix} \varepsilon/2 \\ \varepsilon \end{bmatrix}.$$

Since ε is arbitrary, $S_2(b)$ is empty and β is not identified. ■

The discussion in this section shows that β is identified only in special cases when X is discrete. However, β can be bounded even if it is not identified by using solving the linear programming problem (3.8). This procedure is illustrated in the following example.

Example 3.3: Bounding β when It Is Not Identified

Let $k = 2$ and $\beta = (1, -0.5)'$. Suppose that $P(Y = 1|x)$ and the points of support of X are as shown in Table 3.1. Then β is not identified because X has finite support and $P(X\beta = 0) = 0$. The first component of β is known by scale normalization, so only β_2 needs to be bounded. Problem (3.8) is, therefore,

maximize (minimize): b_2

subject to: $b_2 \leq 0$

$$0.6 + 0.5b_2 \geq 0$$

$$1 + b_2 \geq 0$$

It is clear that the solutions are $-1 \leq b_2 \leq 0$. Thus, although b_2 is not identified, it can be bounded from both above and below. ■

3.3 Estimation

We now turn to the problem of estimating β in the binary-response model (3.1) with the location normalization median$(U|x) = 0$. It is assumed throughout the remainder of this chapter that β is identified and that the distribution of X_1 conditional on $\tilde{X}$ has a density. When it is necessary to specify the conditions insuring identification of β, it will be assumed for convenience that conditions

Table 3.1: An Unidentified Binary-Response Model

(x_1, x_2)	$P(Y\|x)$	$x_1 + b_2x_2$
(1, 0)	0.8	1
(0, 1)	0.3	b_2
(0.6, 0.5)	0.6	$0.6 + 0.5b_2$
(1, 1)	0.7	$1 + b_2$

(a) and (b) of Theorem 3.1 are satisfied. However, all results also hold under the conditions of Corollary 3.1.1.

Although the main task of this section is developing estimators of β, it is also important to consider estimating $P(Y = 1|x)$. This problem is addressed in Section 3.3.1. It turns out that little information about $P(Y = 1|x)$ can be obtained without making assumptions that are considerably stronger than the ones needed to estimate β. Estimators of β are developed in Sections 3.3.2 and 3.3.4.

3.3.1 Estimating $P(Y = 1|x)$

The assumptions needed to identify and, as will be seen in Section 3.3.2, estimate β do not require $P(Y = 1|x)$ to be a continuous function of the continuous components of x. As a result, $P(Y = 1|x)$ is not identified over most of the support of X, and estimation of $P(Y = 1|x)$ under these assumptions is not possible. The only information about $P(Y = 1|x)$ that can be obtained is that implied by knowledge of $x\beta$, which is identified and estimable. This information is

$$P(Y = 1|x) \begin{cases} \geq 0.5 \text{ if } x\beta \geq 0 \\ = 0.5 \text{ if } x\beta = 0 \\ < 0.5 \text{ if } x\beta < 0. \end{cases} \tag{3.12}$$

Thus, the numerical value of $P(Y = 1|x)$ is known if $x\beta = 0$. Otherwise, only bounds on $P(Y = 1|x)$ can be obtained.

Now let b_n be a consistent estimator of β. Let x be fixed, and consider the problem of finding an empirical analog of (3.12). Observe that if $x\beta > 0$, then $P(xb_n > 0) \to 1$ as $n \to \infty$. Conversely, if $x\beta < 0$, then $P(xb_n < 0) \to 1$ as $n \to \infty$. On the other hand, if $x\beta = 0$, then xb_n can be either positive or negative, regardless of the size of n. Therefore, the following estimated bounds on $P(Y = 1|x)$ hold with probability approaching 1 as $n \to \infty$.

$$P_n(Y = 1|x) \begin{cases} \geq 0.5 \text{ if } xb_n \geq 0 \\ \leq 0.5 \text{ if } xb_n \leq 0 \end{cases} \tag{3.13}$$

If it is assumed that $P(Y = 1|x)$ is a continuous function of $x\beta$ in a neighborhood of $x\beta = 0$, then the bounds (3.13) can be sharpened to

$$P_n(Y = 1|x) \begin{cases} \geq 0.5 \text{ if } xb_n > 0 \\ = 0.5 \text{ if } xb_n = 0 \\ \leq 0.5 \text{ if } xb_n < 0. \end{cases} \tag{3.14}$$

Finally, suppose it is assumed that $P(Y = 1|x)$ is a continuous function of the components of x that correspond to continuous components of X. Then $P(Y = 1|x)$ can be estimated as the nonparametric mean-regression of Y on X. The continuity assumption required by this approach is reasonable in many applications but is stronger than necessary to identify or estimate β.

3.3.2 Estimating β: The Maximum Score Estimator

The maximum score estimator was proposed by Manski (1975, 1985) as a method for estimating β consistently in the binary-response model (3.1) when median$(U|x) = 0$. This estimator is the binary-response analog of the least-absolute-deviations (LAD) estimator of a linear median-regression model. In a linear median-regression model, the dependent variable Y is related to explanatory variables X by

$$Y = X\beta + U\ ,$$

where median$(U|x) = 0$. It follows that median$(Y|x) = x\beta$. Moreover, β minimizes the quantity

$$S_{lin}(b) \equiv E|Y - Xb| \tag{3.15}$$

whenever the expectation exists. The LAD estimator of β minimizes the sample analog of $E|Y - Xb|$ that is obtained by replacing the expectation with a sample average. Thus, the LAD estimator solves the problem

$$\text{minimize: } n^{-1}\sum_{i=1}^{n}|Y_i - X_i b|,$$

where $\{Y_i, X_i: i = 1,...,n\}$ is a random sample of $\{Y, X\}$.

To motivate the maximum score estimator, observe that by the definition of the median,

$$\text{median}(Y|x) = \inf\{y: P(Y \geq y|x) \geq 0.5\}\ . \tag{3.16}$$

Recall the definition of the indicator function: $I(\bullet) = 1$ if the event in parentheses occurs and 0 otherwise. Then, combining (3.12) and (3.16) yields

$$\text{median}(Y|x) = I(x\beta \geq 0) .$$

Therefore the binary-response version of (3.15) is

$$(3.17) \qquad S_{bin}(b) = E|Y - I(Xb \geq 0)| .$$

Some easy algebra shows that

$$(3.18) \qquad S_{bin}(b) = E[Y - (2Y-1)I(Xb \geq 0)] .$$

Therefore

$$S_{bin}(b) - S_{bin}(\beta) = E\{(2Y-1)[I(X\beta \geq 0) - I(Xb \geq 0)]\} ,$$

and

$$(3.19) \qquad S_{bin}(b) - S_{bin}(\beta) = E\{[2P(Y=1|x) - 1)][I(X\beta \geq 0) - I(Xb \geq 0)]\} .$$

The right-hand side of (3.19) is zero if $b = \beta$ and non-negative otherwise. Therefore, $b = \beta$ minimizes $S_{bin}(b)$. Indeed, by using arguments similar to those made following the statement of Theorem 3.1, it can be shown that β is the unique minimizer of $S_{bin}(b)$ under the assumptions of that theorem. Thus, we have

Theorem 3.3: Let median$(U|x) = 0$ for all x in the support of X. Let the first component if β equal 1. Then β is the unique minimizer of $E[S_{bin}(b)]$ if

(a) The support of the distribution of X is not contained in any proper linear subspace of $\Re^k$.

(b) For almost every $\tilde{x} = (x_2, \ldots, x_k)$ the distribution of X_1 conditional on $\tilde{X} = \tilde{x}$ has an everywhere positive density. ■

This result is proved formally in Manski (1985).

The sample analog of $E[S_{bin}(b)]$ is obtained by replacing the expectation with a sample average. This yields

$$(3.20) \qquad \tilde{S}_n(b) \equiv n^{-1} \sum_{i=1}^{n} Y_i - n^{-1} \sum_{i=1}^{n} (2Y_i - 1) I(X_i b \geq 0) .$$

The maximum score estimator of β minimizes $\widetilde{S}_n(b)$. Observe, however, that the first term on the right-hand side of (3.20) does not depend on b and, therefore, does not affect the minimization of $\widetilde{S}_n(b)$. Therefore, taking account of the minus sign that precedes the second term, it suffices to solve

$$\text{(3.21)} \qquad \underset{b_1=1}{\text{maximize:}} \; S_{ms}(b) \equiv n^{-1}\sum_{i=1}^{n}(2Y_i - 1)I(X_i b \geq 0)$$

Any solution to (3.21) is a maximum score estimator of β.

Equations (3.17) and (3.18) show that the maximum score estimator is the binary-response analog of the LAD estimator of the coefficient vector of a linear median-regression model.. Another interpretation of the maximum score estimator can be obtained by observing that solving (3.21) is equivalent to solving

$$\text{(3.22)} \qquad \underset{b_1=1}{\text{maximize:}} \; S_{ms}^{*}(b) \equiv \sum_{i=1}^{n}(2Y_i - 1)[2I(X_i b \geq 0) - 1]$$

The summand in (3.22) equals 1 if $Y_i = 1$ and $X_i b \geq 0$ or if $Y_i = 0$ and $X_i b < 0$. The summand equals -1 otherwise. Suppose one predicts that $Y_i = 1$ if $X_i b \geq 0$ and $Y_i = 0$ otherwise. Assign a score value of 1 if the predicted and observed values of Y_i are the same and -1 if they are not. Then $S_{ms}^{*}(b)$ is the sum of the scores, and the maximum score estimator maximizes the total score or total number of correct predictions.

Manski (1985) has proved that the maximum score estimator is strongly consistent for β. This result is stated formally in Theorem 3.4:

Theorem 3.4: Let median$(U|x) = 0$ for all x in the support of X. Let the first component if β equal 1. Assume that there is a known, compact set B that contains β. Let b_n solve

$$\text{(3.23)} \qquad \underset{\substack{b_1=1 \\ b\in B}}{\text{maximize:}} \; S_{ms}^{*}(b) \equiv \sum_{i=1}^{n}(2Y_i - 1)[2I(X_i b \geq 0) - 1]$$

or, equivalently,

$$\text{(3.24)} \qquad \underset{\substack{b_1=1 \\ b\in B}}{\text{maximize:}} \; S_{ms}(b) \equiv n^{-1}\sum_{i=1}^{n}(2Y_i - 1)I(X_i b \geq 0) .$$

Then $b_n \to \beta$ almost surely as $n \to \infty$ if

(a) $\{Y_i, X_i\}$ is a random sample from the distribution of (Y, X).

(b) The support of the distribution of X is not contained in any proper linear subspace of $\Re^k$.

(c) For almost every $\tilde{x} = (x_2, \ldots, x_k)$ the distribution of X_1 conditional on $\tilde{X} = \tilde{x}$ has an everywhere positive density. ■

Cavanagh (1987) and Kim and Pollard (1990) have derived the rate of convergence and asymptotic distribution of the maximum score estimator. This is a difficult task because $S_{ms}(b)$ and $S^*_{ms}(b)$ are discontinuous functions of b. Therefore, the standard Taylor series methods of asymptotic distribution theory cannot be applied to the maximum score estimator. Cavanagh (1987) and Kim and Pollard (1990) show that the maximum score estimator converges in probability at the rate $n^{-1/3}$. The limiting distribution of $n^{1/3}(b_n - \beta)$ is that of the maximum of a multidimensional Brownian motion with quadratic drift. This distribution is too complex for use in carrying out statistical inference in applications.

Manski and Thompson (1986) proposed using the bootstrap to estimate the distribution of the maximum score estimator and to carry out statistical inference. The bootstrap estimates the distribution of a statistic by treating the estimation data as if they were the population. Thus, the estimated distribution of the maximum score estimator is the distribution that is obtained by repeatedly resampling the data randomly with replacement. The resampling procedure can be carried out easily as a Monte Carlo simulation on a microcomputer. Manski and Thompson provided Monte Carlo evidence indicating that the bootstrap gives good estimates of the root mean square error (RMSE) of the maximum score estimator.

Although the RMSE provides a convenient measure of estimation precision, it cannot be used for statistical inference (e.g., obtaining confidence intervals or testing hypotheses about β). This is because the maximum score estimator is not normally distributed, even asymptotically. Therefore, there is no simple relation between the RMSE and the distribution of the estimator. By contrast, the RMSE completely specifies the asymptotic distribution of an estimator that is asymptotically normally distributed. Horowitz (1993a) proposed using the bootstrap to obtain confidence intervals for the components of β. Let b_n^* be the estimate of β that is obtained from a bootstrap sample. Then the probability distribution of $b_n - \beta$ is estimated by the empirical distribution of $b_n^* - b_n$ that is obtained by repeatedly resampling the estimation data. Let b_{ni} and β_i, respectively, denote the i'th components of b_n and β ($i \neq 1$). Let b_{ni}^* denote the i'th component of b_n^*. Finally, let $q_{\alpha i}$ be the $(1 - \alpha)$ quantile of the empirical distribution of $|b_{ni}^* - b_{ni}|$. Then the estimated $100(1 - \alpha)\%$ confidence interval for β_i, is $b_n - q_{\alpha i} \leq \beta_i \leq b_n + q_{\alpha i}$.

The available Monte Carlo evidence suggests that the bootstrap performs well in estimating the distribution of the maximum score estimator. There has been no formal proof, however, that the bootstrap estimator of the distribution of $n^{1/3}(b_n^* - b_n)$ converges to the asymptotic distribution of $n^{1/3}(b_n - \beta)$ as $n \to \infty$.

The rate of convergence of the maximum score estimator, $n^{-1/3}$, is much slower than the $n^{-1/2}$ rate of parametric and single-index estimators. The slow rate of convergence of the maximum score estimator is sometimes interpreted as implying that this estimator is not as useful as parametric or single-index estimators. This interpretation is incorrect. The maximum score estimator makes assumptions about the distribution of (Y, X) that are different from the assumptions made by parametric and single-index estimators. Most importantly for influencing the rate of convergence, the maximum score estimator permits U to have arbitrary heteroskedasticity of unknown form subject to the centering restriction median$(U|x) = 0$. In contrast, single-index models permit only limited forms of heteroskedasticity, and parametric models require the form of any heteroskedasticity to be known up to finitely many parameters. As will be discussed in Section 3.3.3, the maximum score estimator converges at the fastest rate possible under its assumptions. Thus, its slow rate of convergence is not a defect but a reflection of the difficulty of carrying out estimation under its assumptions, especially its assumptions about heteroskedasticity.

3.3.3 Estimating β: The Smoothed Maximum Score Estimator

The maximum score estimator has a complicated limiting distribution that is difficult to derive because the maximum score objective function is discontinuous. Horowitz (1992) proposed smoothing the discontinuities in $S_{ms}^*(b)$ to achieve a differentiable function. He showed that under assumptions that are slightly stronger than those of Theorem 3.4, the resulting smoothed maximum score estimator has a limiting normal distribution and a rate of convergence that is at least $n^{-2/5}$ and can be arbitrarily close to $n^{-1/2}$ under certain smoothness assumptions. This section describes the smoothed maximum score estimator.

The smoothed maximum score estimator can be obtained from either $S_{ms}(b)$ or $S_{ms}^*(b)$, and its properties are the same regardless of which is used. Only $S_{ms}(b)$ will be considered in this discussion. $S_{ms}(b)$ is a discontinuous function of b because each term in the sum on the right-hand side of (3.24) contains the factor $I(X_i b \geq 0)$, which is a step function. The smoothed maximum score estimator is obtained by replacing this indicator function with a function that is twice differentiable. To do this, let K be a function satisfying $|K(v)| < M$ for all v and some $M < \infty$, $\lim_{v \to -\infty} K(v) = 0$, and $\lim_{v \to \infty} K(v) = 1$. K can be thought of as the integral of a kernel function for nonparametric density estimation. K is not a kernel function itself. Let $\{h_n\}$ be a sequence of positive numbers (called bandwidths) that converges to 0 as $n \to \infty$. The smoothed maximum score estimator of β solves

$$\text{(3.25)} \qquad \underset{\substack{b_1=1 \\ b\in B}}{\text{maximize:}} \; S_{sms}(b) \equiv n^{-1}\sum_{i=1}^{n}(2Y_i - 1)K\left(\frac{X_i b}{h_n}\right).$$

Horowitz (1992) showed that the smoothed maximum score estimator is consistent for β under the conditions of Theorem 3.4. The intuition for this result is as follows. It is not difficult to prove that $|S_{ms}(b) - S_{sms}(b)| \to 0$ with probability 1 as $n \to \infty$ uniformly over $b \in B$ (see Horowitz 1992, Lemma 4). The reason for this is that as $h_n \to 0$,

$$K\left(\frac{X_i b}{h_n}\right) \to 1 \;\text{ if } X_i b > 0$$

$$\to 0 \;\text{ if } X_i b < 0$$

for each $i = 1,\ldots, n$. Therefore,

$$K\left(\frac{X_i b}{h_n}\right) \to I(X_i b \geq 0)$$

if $X_i b \neq 0$. $K(X_i b/h_n)$ may not converge to $I(X_i b \geq 0)$ if $X_i b = 0$, but $P(X_i b = 0) = 0$. Therefore, with probability 1, $S_{sms}(b)$ and $S_{ms}(b)$ can be made arbitrarily close to one another for all $b \in B$ by making n sufficiently large. Indeed, it can be shown that this result holds uniformly over $b \in B$. Therefore, the values of b that maximize $S_{sms}(b)$ and $S_{ms}(b)$ can also be made arbitrarily close to one another.

The rate of convergence and asymptotic distribution of the smoothed maximum score estimator can be obtained by using Taylor series methods like those used to obtain the asymptotic distributions of parametric estimators (see, e.g., Amemiya 1985). To do this, define $k \equiv \dim(\beta)$, and let β_j denote the j'th component of β. Let $\tilde{\beta}$ be the $(k - 1)\times 1$ vector consisting of components 2 through k of β. Similarly, let b_n denote any solution to (3.25), and let $\tilde{b}_n$ denote components 2 through k of b_n. Observe that because β_1, the first component of β, can have only two values, every consistent estimator b_{n1} of β_1 satisfies $\lim_{n\to\infty} P(b_{n1} = \beta_1) = 1$ and converges in probability faster than n^{-r} for any $r > 0$. This is not true of the remaining components of β, which can take on a continuum of values. Therefore, it is necessary to derive only the asymptotic distribution of $\tilde{b}_n$.

To do this, let β be an interior point of the parameter set B. Since b_n is consistent for β, b_n is an interior point of B with probability approaching 1 as $n \to \infty$. Therefore, with probability approaching 1, b_n satisfies the first-order condition

$$(3.26) \qquad \frac{\partial S_{sms}(b_n)}{\partial \tilde{b}} = 0.$$

A Taylor series expansion of the left-hand side of (3.26) about $b_n = \beta$ gives

$$(3.27) \qquad \frac{\partial S_{sms}(\beta)}{\partial \tilde{b}} + \frac{\partial^2 S_{sms}(b_n^*)}{\partial \tilde{b}\, \partial \tilde{b}'}(\tilde{b}_n - \tilde{\beta}) = 0,$$

where b_n* is between $\tilde{b}_n$ and $\tilde{\beta}$. Now suppose that there is a non-singular matrix Q such that

$$(3.28) \qquad \operatorname*{plim}_{n\to\infty} \frac{\partial^2 S_{sms}(b_n^*)}{\partial \tilde{b}\, \partial \tilde{b}'} = Q.$$

Suppose, also, that as $n \to \infty$,

$$(3.29) \qquad n^r \frac{\partial S_{sms}(\beta)}{\partial \tilde{b}} \to^d W$$

for some $r > 0$ and some random variable W. Then (3.27) can be written in the form

$$W + Qn^r(\tilde{b}_n - \tilde{\beta}) = o_p(1)$$

and

$$n^r(\tilde{b}_n - \tilde{\beta}) = -Q^{-1}W + o_p(1).$$

Thus, $n^r(\tilde{b}_n - \tilde{\beta})$ is asymptotically distributed as $-Q^{-1}W$. It turns out that W is normally distributed, as is discussed below. The value of r and, therefore, the rate of convergence of the smoothed maximum score estimator depend on certain smoothness properties of $P(Y = 1|x)$ and the probability distribution of

$X\beta$ conditional on $\widetilde{X}$. Under regularity conditions that are given below, the rate of convergence is at least $n^{-2/5}$ and can be arbitrarily close (though not equal to) $n^{-1/2}$ if $P(Y = 1|x)$ and the conditional distribution of $X\beta$ are sufficiently smooth.

Additional notation is needed to formalize these ideas and make them precise. Define $Z = X\beta$ and observe that since $|\beta_1| = 1$, there is a one-to-one relation between $(Z, \widetilde{X})$ and X for any fixed β. Assume that the distribution of Z conditional on $\widetilde{X} = \widetilde{x}$ has a probability density, $p(z|\widetilde{x})$ that is strictly positive for almost every $\widetilde{x}$. For each positive integer j, define

$$p^{(j)}(z|\widetilde{x}) = \frac{\partial^j p(z|\widetilde{x})}{\partial z^j})$$

whenever the derivative exists, and define $p^{(0)}(z|\widetilde{x}) = p(z|\widetilde{x})$. Let P denote the CDF of $\widetilde{X}$, and let $F(\bullet|z, \widetilde{x})$ denote the CDF of U conditional on $Z = z$ and $\widetilde{X} = \widetilde{x}$. For each positive integer j, define

$$F^{(j)}(-z|z,\widetilde{x}) = \frac{\partial^j F(-z|z,\widetilde{x})}{\partial z^j}$$

whenever the derivative exists. For example,

$$F^{(1)}(-z|z,\widetilde{x}) = \left[\frac{\partial F(u|z,\widetilde{x})}{\partial u} - \frac{\partial F(u|z,\widetilde{x})}{\partial z}\right]_{u=-z}.$$

Let K' and K'', respectively, denote the first and second derivatives of K. Let s be a positive integer that is defined according to criteria that are stated below. Define the scalar constants α_A and α_D by

$$\alpha_A = \int_{-\infty}^{\infty} v^s K'(v)dv$$

and

$$\alpha_D = \int_{-\infty}^{\infty} [K'(v)^2]dv$$

whenever these quantities exist. Define the $(k - 1)\times 1$ vector A and the $(k - 1)\times(k - 1)$ matrices D and Q by

$$A = -2\alpha_A \sum_{j=1}^{s} \frac{1}{j!(s-j)!} E[F^{(j)}(0|0,\widetilde{X}) p^{(s-j)}(0|\widetilde{X})\widetilde{X}'],$$

$$D = \alpha_D E[\widetilde{X}'\widetilde{X} p(0|\widetilde{X})],$$

and

$$Q = 2E[\widetilde{X}'\widetilde{X} F^{(1)}(0|0,\widetilde{X}) p(0|\widetilde{X})]$$

whenever these quantities exist.

The asymptotic distribution of the smoothed maximum score estimator is derived under the following assumptions:

Assumption 1: $\{Y_i, X_i: i = 1, ..., n\}$ is a random sample of (Y, X), where $Y = I(X\beta + U \geq 0)$, $X \in \Re^k$ $(k \geq 1)$, U is a random scalar, and $\beta \in \Re^k$ is a constant.

Assumption 2: (a) The support of the distribution of X is not contained in any proper linear subspace of $\Re^k$. (b) $0 < P(Y = 1|x) < 1$ for almost every x. (c) $|\beta_1| = 1$, and for almost every $\widetilde{x} \equiv (x_2, ..., x_k)$, the distribution of X_1 conditional on $\widetilde{X} = \widetilde{x}$ has an everywhere positive probability density.

Assumption 3: Median$(U|x) = 0$ for almost every x.

Assumption 4: $\widetilde{\beta} \equiv (\beta_2, ..., \beta_k)'$ is an interior point of a compact set $\widetilde{B} \in \Re^{k-1}$.

Assumption 5: The components of $\widetilde{X}$ and of the matrices $\widetilde{X}'\widetilde{X}$ and $\widetilde{X}'\widetilde{X}\widetilde{X}'\widetilde{X}$ have finite first absolute moments.

Assumption 6: As $n \to \infty$, $h_n \to 0$ and $(\log n)/(nh_n^4) \to 0$.

Assumption 7: (a) K is twice differentiable everywhere, $|K'(\bullet)|$ and $|K''(\bullet)|$ are uniformly bounded, and each of the following integrals over $(-\infty, \infty)$ is finite:

$$\int [K'(v)]^4 dv, \quad \int [K''(v)]^2 dv, \quad \int |v^2 K''(v)| dv.$$

(b) For some integer $s \geq 2$ and each integer j $(1 \leq j \leq s)$

$$\int_{-\infty}^{\infty} v^j K'(v)dv = 0 \quad \text{if } j < s$$

(3.30)

$$= d \quad \text{(nonzero) if } j = s.$$

(c) For each integer j between 0 and s, any $\eta > 0$, and any sequence $\{h_n\}$ converging to 0

$$\lim_{n \to \infty} h_n^{j-s} \int_{|h_n v| > \eta} |v^j K'(v)| dv = 0$$

and

$$\lim_{n \to \infty} h_n^{-1} \int_{|h_n v| > \eta} |K''(v)| dv = 0 .$$

Assumption 8: For each integer j such that $1 \leq j \leq s - 1$, all z in a neighborhood of 0, almost every $\tilde{x}$, and some $M < \infty$, $p^{(j)}(z|\tilde{x})$ exists and is a continuous function of z satisfying $|p^{(j)}(z|\tilde{x})| < M$ for all z and almost every $\tilde{x}$.

Assumption 9: For each integer j such that $1 \leq j \leq s$, all z in a neighborhood of 0, almost every $\tilde{x}$, and some $M < \infty$, $F^{(j)}(-z|z, \tilde{x})$ exists and is a continuous function of z satisfying $|F^{(j)}(-z|z, \tilde{x})| < M$.

Assumption 9 is satisfied if $[\partial^{j+k} F(u|z, \tilde{x})/\partial u^j \partial z^k]_{u=-z}$ is bounded and continuous in a neighborhood of $z = 0$ for almost every $\tilde{x}$ whenever $j + k \leq s$.

Assumption 10: The matrix Q is negative definite.

The reasons for making these assumptions will now be explained. Assumptions 1-3 except for 2(b) and the compactness part of assumption 4 are the requirements for consistency of both the smoothed and the unsmoothed maximum score estimators. Assumption 2b rules out certain degenerate cases in which β can be learned perfectly from a finite sample (Manski 1985). The assumption that $\tilde{\beta}$ is an interior point of the parameter set insures that the first-order condition $\partial S_{sms}(b_n)/\partial \tilde{b} = 0$ is satisfied with probability approaching 1 as $n \to \infty$. A similar assumption is made for the same reason in parametric maximum likelihood estimation. Assumptions 5 and 7-9 insure the existence of A, D, and Q as well as the convergence of certain sequences that arise in the proof of asymptotic normality. Examples of functions satisfying assumption 7 are given by Müller (1984).

Assumptions 6-9 are analogous to assumptions made in kernel density estimation. In kernel density estimation, a kernel K' that satisfies (3.30) is an

s'th order kernel. With an s'th order kernel and a bandwidth parameter h_n satisfying $nh_n \to \infty$, the bias of the density estimator is $O(h_n^s)$, and the variance is $O[(nh_n)^{-1}]$ if the density being estimated is s times differentiable. Moreover, with an s'th order kernel, the fastest achievable rate of convergence of the density estimator is $n^{-s/(2s+1)}$, so use of a higher-order kernel speeds convergence if the required derivatives of the density exist. Analogous results hold in smoothed maximum score estimation. If assumptions 1-10 are satisfied, the bias of the smoothed maximum score estimator is $O(h_n^s)$, the variance is $O[(nh_n)^{-1}]$, and the fastest achievable rate of convergence is $n^{-s/(2s+1)}$. Thus, faster convergence can be achieved by using a higher order K' if the necessary derivatives of F and p exist.

The matrix Q is analogous to the Hessian form of the information matrix in parametric quasi maximum likelihood estimation (White 1982), and assumption 10 is analogous to the familiar assumption that the Hessian information matrix is nonsingular.

The main results concerning the asymptotic distribution of the smoothed maximum score estimator are given by the following theorem.

Theorem 3.5: Let assumptions 1-10 hold for some $s \geq 2$, and let $\{b_n\}$ be a sequence of solutions to problem (3.25). Then:

(a) If $nh_n^{2s+1} \to \infty$ as $n \to \infty$, $h_n^{-s}(\tilde{b}_n - \tilde{\beta}) \to^p -Q^{-1}A$.

(b) If nh_n^{2s+1} has a finite limit λ as $n \to \infty$,

$$(nh_n)^{1/2}(\tilde{b}_n - \tilde{\beta}) \to^d N(-\lambda^{1/2}Q^{-1}A, Q^{-1}DQ^{-1}).$$

(c) Let $h_n = (\lambda/n)^{1/(2s+1)}$ with $0 < \lambda < \infty$. Let Ω be any nonstochastic, positive semidefinite matrix such that $A'Q^{-1}\Omega Q^{-1}A \neq 0$. Let E_A denote the expectation with respect to the asymptotic distribution of $(nh_n)^{1/2}(\tilde{b}_n - \tilde{\beta})$, and $MSE \equiv E_A(\tilde{b}_n - \tilde{\beta})'\Omega(\tilde{b}_n - \tilde{\beta})$. MSE is minimized by setting

$$(3.31) \qquad \lambda^* \equiv \frac{\text{trace}\,(Q^{-1}\Omega Q^{-1}D)}{2sA'Q^{-1}\Omega Q^{-1}A},$$

in which case

$$n^{s/(2s+1)}(\tilde{b}_n - \tilde{\beta}) \to^d N[-(\lambda^*)^{s/(2s+1)}Q^{-1}A, (\lambda^*)^{-1/(2s+1)}Q^{-1}DQ^{-1}]. \blacksquare$$

Theorem 3.5 implies that the fastest possible rate of convergence in probability of $\tilde{b}_n$ is $n^{-s/(2s+1)}$. A sufficient condition for this to occur is

$$h_n = \left(\frac{\lambda}{n}\right)^{1/(2s+1)}$$

with $0 < \lambda < \infty$. The asymptotically optimal value of λ in the sense of minimizing *MSE* is λ^*. Horowitz (1993c) shows that $n^{-s/(2s+1)}$ is the fastest achievable rate of convergence of any estimator of $\tilde{\beta}$ under assumptions 1-5 and 8-10. Thus, no estimator can converge faster than the smoothed maximum score estimator under these assumptions. The fastest rate of convergence if $s < 2$ is $n^{-1/3}$. Moreover, $n^{1/3}(\tilde{b}_n - \tilde{\beta})$ has a complicated, non-normal limiting distribution if $s < 2$. Thus, in terms of the rate of convergence and simplicity of the asymptotic distribution, the smoothed maximum score estimator has no advantage over the unsmoothed one unless $F(-z|z, \tilde{x})$ and $p(z|\tilde{x})$ have sufficiently many derivatives.

Theorem 3.5 also implies that the mean of the asymptotic distribution of $n^{s/(2s+1)}(\tilde{b}_n - \tilde{\beta})$ is not 0 when $h_n \propto n^{-s/(2s+1)}$. In other words, $n^{s/(2s+1)}(\tilde{b}_n - \tilde{\beta})$ is asymptotically biased when h_n is chosen so that $\tilde{b}_n$ has its fastest possible rate of convergence. Asymptotic bias also arises in kernel nonparametric density estimation and nonparametric mean-regression. One way to remove the asymptotic bias is by using a bandwidth h_n that converges more rapidly than $n^{-1/(2s+1)}$. This is called *undersmoothing*. Theorem 3.5(b) shows that $\lambda = 0$ with undersmoothing, so there is no asymptotic bias.

Undersmoothing as a method of bias reduction has the disadvantage that it slows the rate of convergence of $\tilde{b}_n$ to $\tilde{\beta}$. This problem can be avoided by forming an estimate of the bias and subtracting the estimate from $\tilde{b}_n$. A method for doing this is described below.

To make the results of Theorem 3.5 useful in applications, it is necessary to be able to estimate A, D, and Q consistently. Theorem 3.6 shows how this can be done. For any $b \in B$ and $h > 0$, define

$$D_n = \frac{1}{nh_n}\sum_{i=1}^{n} \tilde{X}_i'\tilde{X}_i\left[K'\left(\frac{X_i b_n}{h_n}\right)\right]^2$$

and

$$Q_n = \frac{\partial^2 S_{sms}(b_n)}{\partial \tilde{b}\, \partial \tilde{b}'}.$$

Theorem 3.6: Let b_n be a consistent smoothed maximum score estimator based on the bandwidth $h_n \propto n^{-\kappa}$, where $\kappa \geq 1/(2s+1)$. Then $D_n(b_n) \to^p D$ and $Q_n \to^p Q$ as $n \to \infty$. Let $\{h_n^*\}$ be a sequence of positive numbers satisfying $h_n^* \propto n^{-\delta/(2s+1)}$, where $0 < \delta < 1$. Define

$$A_n = \frac{1}{n\left(h_n^*\right)^{s+1}} \sum_{i=1}^{n} (2Y_i - 1)\tilde{X}_i' K'\left(\frac{X_i b_n}{h_n^*}\right).$$

If $h_n \propto n^{-1/(2s+1)}$, then $A_n \to^p A$ as $n \to \infty$. ■

Theorem 3.6 implies that the covariance matrix of the asymptotic distribution of $(nh_n)^{1/2}(\tilde{b}_n - \tilde{\beta})$, $Q^{-1}DQ^{-1}$, is estimated consistently by

$$(3.32) \quad V_n = Q_n^{-1} D_n Q_n^{-1}.$$

This result will be used below to form tests of hypotheses about $\tilde{\beta}$.

Theorem 3.6 also provides a way to remove the asymptotic bias of $\tilde{b}_n$ without slowing its rate of convergence. Recall that that if $h_n \propto n^{-1/(2s+1)}$, then the asymptotic bias of $(nh_n)^{1/2}(\tilde{b}_n - \tilde{\beta})$ is $-\lambda^{s/(2s+1)}Q^{-1}A$. By Theorem 3.6, this bias is estimated consistently by $-\lambda^{s/(2s+1)}Q_n^{-1}A_n$. Therefore, an asymptotically unbiased estimator of $\tilde{\beta}$ is

$$\hat{b}_n = \tilde{b}_n + \left(\frac{\lambda}{n}\right)^{s/(2s+1)} Q_n^{-1} A_n.$$

The asymptotic normality of $(nh_n)^{1/2}(\tilde{b}_n - \tilde{\beta})$ makes it possible to form asymptotic t and χ^2 statistics for testing hypotheses about $\tilde{\beta}$. Consider, first, a t test of a hypothesis about a component of $\tilde{\beta}$. Let $\tilde{b}_{nj}$, $\hat{b}_{nj}$, and $\tilde{\beta}_j$, respectively, denote the j'th components of $\tilde{b}_n$, $\hat{b}_n$, and $\tilde{\beta}$. Let V_{nj} denote the (j, j) component of the matrix V_n defined in (3.32). The t statistic for testing the hypothesis H_0: $\tilde{\beta}_j = \tilde{\beta}_{j0}$ is

$$t = \frac{(nh_n)^{1/2}(\hat{b}_{nj} - \tilde{\beta}_{j0})}{V_{nj}^{1/2}}$$

if $h_n \propto n^{-1/(2s+1)}$ and

$$t = \frac{(nh_n)^{1/2}(\tilde{b}_{nj} - \tilde{\beta}_{j0})}{V_{nj}^{1/2}}$$

with undersmoothing. In either case, t is asymptotically distributed as $N(0,1)$ if H_0 is true. Therefore, H_0 can be rejected or accepted by comparing t with the relevant quantile of the standard normal distribution.

Now let R be an $r\times(k - 1)$ matrix with $r \le k - 1$, and let c be an $r\times 1$ vector of constants. Consider a test of the hypothesis H_0: $R\tilde{\beta} = c$. Assume that the matrix $RQ^{-1}DQ^{-1}R'$ is nonsingular. Suppose that $h_n \propto n^{-1/(2s+1)}$. Then under H_0, the statistic

$$\chi^2 = (nh_n)(R\hat{b}_n - c)'(RV_nR')^{-1}(R\hat{b}_n - c)$$

is asymptotically chi-square distributed with r degrees of freedom. If undersmoothing is used so that $h_n \propto n^{-\kappa}$ with $\kappa > 1/(2s + 1)$, then the statistic

$$\chi^2 = (nh_n)(R\tilde{b}_n - c)'(RV_nR')^{-1}(R\tilde{b}_n - c)$$

is asymptotically chi-square distributed with r degrees of freedom. In either case, H_0 can be accepted or rejected by comparing χ^2 with the appropriate quantile of the chi-square distribution with r degrees of freedom.

The asymptotic distributions of t and χ^2 are only approximations to the exact, finite-sample distributions of these statistics. The approximation errors can be made arbitrarily small by making n sufficiently large. With the sample sizes encountered in applications, however, the approximation errors can be large. As a result, the true and nominal probabilities of rejecting a correct null hypothesis can be very different when the critical value is obtained from the asymptotic distribution of the test statistic. For example, a symmetrical t test rejects a true null hypothesis with nominal probability α if $|t|$ exceeds the $(1 - \alpha)$ quantile of the standard normal distribution. However, the true rejection probability may be much larger than α if, as often happens in finite samples, the asymptotic distribution of t (the standard normal distribution) is not a good approximation to its exact distribution. Horowitz (1992) provides Monte Carlo evidence showing that the true rejection probability can exceed the nominal probability by a factor of three or more with samples of practical size.

This problem can be greatly reduced through the use of the bootstrap. As was discussed in Section 3.3.2, the bootstrap estimates the distribution of a statistic by treating the estimation data as if they were the population. The bootstrap distribution of a statistic is the distribution induced by sampling the estimation data randomly with replacement. The α-level bootstrap critical value of a symmetrical t test is the $1 - \alpha$ quantile of the bootstrap distribution of $|t|$. The α-level bootstrap critical value of a test based on χ^2 is the $1 - \alpha$ quantile of the bootstrap distribution of χ^2. Similar procedures can be used to obtain bootstrap critical values for one-tailed and equal-tailed t tests.

Under certain conditions (see, e.g., Beran 1988; Hall 1986, 1992), the bootstrap provides a better finite-sample approximation to the distribution of a statistic than does asymptotic distribution theory. When this happens, the use of bootstrap critical values instead of asymptotic ones reduces the differences between the true and nominal probabilities that a t or χ^2 test rejects a true null hypothesis. The use of bootstrap critical values also reduces the differences between the true and nominal coverage probabilities of confidence intervals. To achieve these results, the statistic in question must be *asymptotically pivotal*, meaning that its asymptotic distribution does not depend on unknown population parameters. Horowitz (1997) reviews the theory of the bootstrap and provides numerical examples of its use in econometrics.

The t and χ^2 statistics for testing hypotheses using smoothed maximum score estimates are asymptotically pivotal. However, they do not satisfy the standard regularity conditions under which the bootstrap provides asymptotic refinements (that is, improvements in the approximation to the finite-sample distribution of a statistic). Nonetheless, it is possible to prove that the bootstrap provides asymptotic refinements for t and χ^2 tests based on the smoothed maximum score estimator. See Horowitz (1996a). The main problem is that the standard theory of the bootstrap assumes that the statistic in question can be approximated by a function of sample moments whose probability distribution has an Edgeworth expansion. However, because the t and χ^2 statistics based on the smoothed maximum score estimator depend on the bandwidth parameter h_n that decreases to 0 as $n \to \infty$, they cannot be approximated by functions of sample moments. Horowitz (1996a) shows how to modify the standard theory of Edgeworth expansions to deal with this problem.

The bootstrap distributions of t and χ^2 cannot be calculated analytically, but they can be estimated with arbitrary accuracy by Monte Carlo simulation. To specify the Monte Carlo procedure, let the bootstrap sample be denoted by $\{Y_i^*, X_i^*: i = 1, ..., n\}$. This sample is obtained by sampling the estimation data randomly with replacement. Define the following bootstrap analogs of $S_{sms}(b)$, $Q_n(b)$, and $D_n(b)$:

$$S_{sms}^*(b) \equiv n^{-1} \sum_{i=1}^{n} (2Y_i^* - 1) K\left(\frac{X_i^* b}{h_n}\right),$$

$$D_n^*(b) = \frac{1}{nh_n} \sum_{i=1}^{n} \widetilde{X}_i^{*\prime} \widetilde{X}_i^* \left[K'\left(\frac{X_i^* b}{h_n} \right) \right]^2 ,$$

and

$$Q_n(b) = \frac{\partial^2 S_{sms}^*(b)}{\partial \widetilde{b} \, \partial \widetilde{b}'} .$$

Let $b_n^* = (b_{n1}^*, \widetilde{b}_n^*)$ be a solution to (3.25) with S_{sms} replaced by S_{sms}^*. Let V_{nj} be the (j, j) component of the matrix $Q_n^*(b_n^*)^{-1} D_n^*(b_n^*)\, Q_n^*(b_n^*)^{-1}$.

The Monte Carlo procedure for estimating the bootstrap critical value of the symmetrical t test is as follows. The procedures for estimating bootstrap critical values of one-tailed and equal-tailed t tests and the χ^2 test are similar.

1. Generate a bootstrap sample $\{Y_i^*, X_i^*: \ i = 1, ..., n\}$ by sampling the estimation data randomly with replacement.

2. Using the bootstrap sample, compute the bootstrap t statistic for testing the bootstrap hypothesis H_0^*: $\widetilde{\beta}_j = \widetilde{b}_{nj}$, where b_n solves (3.25). The bootstrap t statistic is

$$t^* = \frac{(nh_n)^{1/2} (\widetilde{b}_{nj}^* - \widetilde{b}_n^j)}{(V_{nj}^*)^{1/2}} ,$$

where $\widetilde{b}_{nj}^*$ is the j'th component of $\widetilde{b}_n^*$.

3. Estimate the bootstrap distribution of $|t^*|$ by the empirical distribution that is obtained by repeating steps 1 and 2 many times. The bootstrap critical value of the symmetrical t test is estimated by the $1 - \alpha$ quantile of this empirical distribution.

The theory of the bootstrap requires h_n to be chosen so as to undersmooth. Thus, it is not necessary to use $\hat{b}_n$ to carry out bootstrap-based testing. Horowitz (1996a) shows that when critical values based on the bootstrap are used for a symmetrical t or χ^2 test, the difference between the true and nominal probabilities of rejecting a true null hypothesis (the error in the rejection probability) has size $o[1/(nh_n)]$. By contrast, the error in the rejection probability exceeds $O[1/(nh_n)]$ if asymptotic critical values are used. Therefore, the error in the rejection probability converges to 0 as $n \to \infty$ more rapidly with bootstrap critical values than with asymptotic ones. Accordingly, the error in the rejection probability is smaller with bootstrap critical values than with asymptotic ones if n is sufficiently large. Similar results are available for one-tailed and equal-tailed t tests, although the order of the asymptotic refinement for these is not the same as it is for symmetrical tests. Horowitz (1996a) presents Monte Carlo evidence indicating that with samples of practical size, the bootstrap essentially eliminates errors in the rejection probabilities of symmetrical t tests. Thus, it is better in applications to use the bootstrap to obtain critical values for a test based on the smoothed maximum score estimator than to use the asymptotic distribution of the test statistic.

A final problem in practical implementation of the smoothed maximum score estimator or tests based on it is bandwidth selection. The bandwidth that minimizes the asymptotic mean-square error of $\tilde{b}_n$ is

$$h_{n,opt} = \left(\frac{\lambda^*}{n}\right)^{1/(2s+1)},$$

where λ^* is given by (3.31). The value of λ^* can be estimated by replacing unknown quantities on the right-hand side of (3.31) by estimates based on a preliminary bandwidth. This is called the *plug-in* method of bandwidth selection. To implement the plug-in method, choose a preliminary bandwidth $h_{n1} \propto n^{-s/(2s+1)}$ and any $h_{n1}^* \propto n^{-\delta/(2s+1)}$ for $0 < \delta < 1$. Compute the smoothed maximum score estimate b_n base on h_{n1}, and use b_n and h_{n1}^* to compute A_n, D_n, and Q_n. Then estimate λ^* by λ_n, where λ_n is obtained from (3.31) by replacing A, D, and Q with A_n, D_n, and Q_n.

This method of bandwidth selection is analogous to the plug-in method of kernel density estimation. As in density estimation, it has the disadvantage of not being fully automatic; the estimated optimal bandwidth depends on the bandwidth used to obtain the initial estimate of β and on δ. Jones, *et al.* (1996) discuss ways to reduce the dependence on the initial bandwidth in the case of density estimation. There has been no research on whether similar methods can be developed for smoothed maximum score estimation.

The bootstrap method for obtaining improved finite-sample critical values for t and χ^2 tests if hypotheses about β requires a bandwidth that undersmooths. That is, it requires h_n to converge faster than $n^{-1/(2s+1)}$. Existing theory provides no guidance on how such a bandwidth should be chosen in applications. However, Monte Carlo evidence presented by Horowitz (1996a) indicates that the empirical levels of tests with bootstrap critical values are not sensitive to the value of h_n over the range $0.5h_{n,opt}$ to $h_{n,opt}$. These results suggest that, as a practical rule-of-thumb method for selecting the bandwidth for bootstrap calculations, one can use the plug-in method to estimate $h_{n,opt}$ and then implement the bootstrap with a bandwidth that is between the estimated $h_{n,opt}$ and half the estimate.

3.4 Extensions of the Maximum Score and Smoothed Maximum Score Estimators

This section shows how the maximum score and smoothed maximum score estimators can be extended for use with choice-based samples, panel data, and ordered-response models.

3.4.1 Choice-Based Samples

The discussion in the preceding sections assumes that the estimation data are a random sample from the distribution of (Y, X). A choice based sample is not random in this way. Instead, it is stratified on the dependent variable Y. The fraction of observations with $Y = 1$ is selected by design, and X is sampled randomly conditional on Y. For example, suppose the modes that are available for travel between two cities are airplane and automobile. Then a data set for analyzing the mode choices of travelers between these cities might be obtained by interviewing randomly selected air travelers at the airport and randomly selected automobile travelers at the roadside. Choice based sampling can be much more efficient than random sampling. For example, suppose that one percent of the population makes the choice of interest in a given time period. Then data acquisition by randomly sampling the population would require contacting 100 persons on average to obtain one useful observation. By contrast, all contacts made in a choice-based sample are potentially useful.

Except in special cases, estimators that work with random samples are inconsistent when the sample is choice based. Parametric estimation with choice-based samples has been investigated by Cosslett (1981), Imbens (1992), Hsieh *et al.* (1985), Manski and Lerman (1977), and Manski and McFadden (1981). This section discusses semiparametric estimation with choice-based samples under the assumption that the distribution of U is unknown. It is assumed that the population values of the aggregate shares, $\pi_1 \equiv \Pr(Y = 1)$ and $\pi_0 \equiv 1 - \pi_1$ are known. It is not unusual in applications for aggregate shares to known quite accurately. For example, in the United States, aggregate shares for certain types of travel choices are available from the U. S. Census. However, census data do not include information on all of the variables needed to develop a useful choice model.

Manski (1986) has given conditions under which β can be estimated consistently from a choice-based sample by maximizing the modified maximum-score objective function

$$S_{n,CB}(b) = \frac{\pi_1}{n_1} \sum_{\substack{i=1 \\ Y_i=1}}^{n} I(X_i b \geq 0) - \frac{\pi_0}{n_0} \sum_{\substack{i=1 \\ Y_i=0}}^{n} I(X_i b \geq 0), \tag{3.33}$$

subject to scale normalization, where n_j ($j = 0, 1$) is the number of observations for which $Y_j = j$ and $n = n_0 + n_1$. The modified maximum score estimator solves

$$\underset{\substack{b_1=1 \\ b \in B}}{\text{maximize}}: S_{n,CB}(b). \tag{3.34}$$

The solution to (3.34) is consistent for β under conditions that are stated in the following theorem, which is a modified version of Theorem 3.4. Define k= dim(X).

Theorem 3.7: Let median($U|x$) = 0 for all x in the support of X. Let the first component of β equal 1. Assume that B is a compact subset of $\Re^k$. Let b_n solve problem (3.34). Then $b_n \to \beta$ almost surely as $n_1 \to \infty$ and $n_0 \to \infty$ if

(a) For each j = 0 or 1, $\{X_i: i = 1, ..., n_j\}$ is a random sample from the distribution of X conditional on $Y = j$.

(b) The support of the distribution of X is not contained in any proper linear subspace of $\Re^k$.

(c) For almost every $\tilde{x} = (x_2, \ldots, x_k)$ the distribution of X_1 conditional on $\tilde{X} = \tilde{x}$ has an everywhere positive density. ■

Conditions (b) and (c) of Theorem 3.7 are identical to conditions in Theorem 3.4. Condition (a) of Theorem 3.7 specifies a choice-based sampling process instead of the random sampling process that is used in Theorem 3.4.

To understand why solving (3.34) yields a consistent estimator of β, define

$$q_1 = \frac{n_1}{n}$$

and

$$q_0 = \frac{n_0}{n},$$

respectively, to be the fractions of the sample for which $Y = 1$ and 0. To minimize the complexity of the discussion, suppose that X has a probability density function f. Similar but notationally more complex arguments apply if X has one or more discrete components. Let $f(\bullet|Y = j)$ be the probability density of X conditional on $Y = j$ ($j = 0$ or 1). Observe that by the algebra of conditional probabilities

$$E[I(Xb \geq 0 \mid Y = 1)] = \int I(xb \geq 0) f(x \mid Y = 1) dx$$

$$= \int I(xb \geq 0) \frac{P(Y = 1 \mid x) f(x)}{q_1} dx$$

$$= \frac{1}{q_1} \int P(Y = 1 \mid x) I(xb \geq 0) f(x) dx.$$

Similarly

$$E[I(Xb \geq 0) \mid Y = 0] = \frac{1}{q_0} \int P(Y = 0 \mid x) I(xb \geq 0) f(x) dx.$$

Therefore, the population version of (3.33) is

$$E[S_{n,CB}(b)] = \int [P(Y = 1 \mid x) - P(Y = 0 \mid x)] I(xb \geq 0) f(x) dx$$

$$= E[2P(Y = 1 \mid x) - 1] I(xb \geq 0)]$$

$$= E[Y - S_{bin}(b)],$$

where S_{bin} is as defined in (3.18). It was shown in Section 3.3.2 that β minimizes $E[S_{bin}(b)]$ and, therefore, maximizes $E[Y - S_{bin}(b)]$. It follows that β maximizes the population version of $S_{n,CB}(b)$. Arguments like those used to prove Theorem 3.4 now show that the solution to (3.34) is consistent for β under choice based sampling. Manski (1986) provides the technical details.

$S_{n.CB},(b)$, like $S_{ms}(b)$, is discontinuous and can be smoothed by replacing the indicator function with an integral of a kernel function. The resulting smoothed maximum score estimator for choice-based samples maximizes the objective function

(3.35) $$S_{sms,CB}(b) = \frac{\pi_1}{n_1} \sum_{\substack{i=1 \\ Y_i=1}}^{n} K\left(\frac{X_i b}{h_n}\right) - \frac{\pi_0}{n_0} \sum_{\substack{i=1 \\ Y_i=0}}^{n} K\left(\frac{X_i b}{h_n}\right),$$

where K and h_n are defined as in (3.25). Define s as in Theorem 3.5. Define $Z = X\beta$, and let $p(\bullet|\tilde{x}, Y = j)$ denote the probability density function of Z conditional on $\tilde{X} = \tilde{x}$ and $Y = j$ (j = 1 or 0). Set

$$p^{(s)}(z \mid \tilde{x}, Y = j) = \frac{\partial^s p(z \mid \tilde{x}, Y = j)}{\partial z^s}$$

Suppose that there is a finite λ such that $(nh_n)^{2s+1} \to \lambda$ as $n \to \infty$. By using Taylor series methods like those used to prove Theorem 3.5, it can be shown that under suitable assumptions, the centered, normalized, smoothed maximum score estimator for choice-based samples is asymptotically normally distributed. To state the result formally, define

$$A_C = \frac{\alpha_A}{s!}\{\pi_1 E[\tilde{X}p^{(s)}(0|\tilde{X}, Y = 1)|Y = 1]$$

$$- \pi_0 E[\tilde{X}p^{(s)}(0|\tilde{X}, Y = 0)|Y = 0]\},$$

$$D_C = \frac{\alpha_D}{2}\left(\frac{\pi_1}{q_1} + \frac{\pi_0}{q_0}\right) E[\tilde{X}'\tilde{X}p(0|\tilde{X})],$$

and

$$Q_C = -\pi_1 E[\tilde{X}'\tilde{X}p^{(1)}(0 \mid \tilde{X}, Y = 1) \mid Y = 1]$$

$$+ \pi_0 E[\tilde{X}'\tilde{X}p^{(1)}(0 \mid \tilde{X}, Y = 0) \mid Y = 0].$$

Make the following assumptions.

Assumption CB1: For each j = 0 or 1, $\{X_i: i = 1, ..., n_j\}$ is a random sample from the distribution of X conditional on $Y = j$.

Assumption CB2: (a) The support of the distribution of X is not contained in any proper linear subspace of $\Re^k$. (b) $|\beta_1| = 1$, and for almost every $\tilde{x} \equiv (x_2,$

..., x_k), the distribution of X_1 conditional on $\tilde{X} = \tilde{x}$ has an everywhere positive density.

Assumption CB3: Median($U|x$) = 0 for almost every x.

Assumption CB4: $\tilde{\beta} \equiv (\beta_2, ..., \beta_k)'$ is an interior point of a compact subset $\tilde{B}$ of $\Re^{k-1}$.

Assumption CB5: The components of $\tilde{X}$ and of the matrices $\tilde{X}'\tilde{X}$ and $\tilde{X}'\tilde{X}\tilde{X}'\tilde{X}$ have finite first absolute moments.

Assumption CB6: (a) K is twice differentiable everywhere, $|K'(\bullet)|$ and $|K''(\bullet)|$ are uniformly bounded, and each of the following integrals over $(-\infty, \infty)$ is finite:

$$\int [K'(v)]^4 dv, \quad \int [K''(v)]^2 dv, \quad \int |v^2 K''(v)| dv .$$

(b) For some integer $s \geq 2$ and each integer j $(1 \leq j \leq s)$

$$\int_{-\infty}^{\infty} v^j K'(v) dv = 0 \quad \text{if } j < s$$

$$= d \text{ (nonzero) if } j = s.$$

(c) For each integer j between 0 and s, any $\eta > 0$, and any sequence $\{h_n\}$ converging to 0

$$\lim_{n \to \infty} h_n^{j-s} \int_{|h_n v| > \eta} |v^j K'(v)| dv = 0$$

and

$$\lim_{n \to \infty} h_n^{-1} \int_{|h_n v| > \eta} |K''(v)| dv = 0 .$$

Assumption CB7: For each integer j such that $1 \leq j \leq s$, all z in a neighborhood of 0, almost every $\tilde{x}$, and some $M < \infty$, $p^{(j)}(z|\tilde{x}, Y = m)$ $(m = 0, 1)$ exists and is a continuous function of z satisfying $|p^{(j)}(z|\tilde{x}, Y = m)| < M$ for all z and almost every $\tilde{x}$.

Assumption CB8: The matrix Q_C is negative definite.

Asymptotic normality of the smoothed maximum score estimator for choice-based samples is given by the following theorem.

Theorem 3.8: Let assumptions CB1-CB8 hold for some $s \geq 2$, and let $\{b_n\}$ be a sequence of solutions to problem (3.35). Assume that $nh_n^{2s+1} = \lambda$ for some finite $\lambda > 0$. Then:

$$(3.36) \qquad (nh_n)^{1/2}(\tilde{b}_n - \tilde{\beta}) \rightarrow^d N(-\lambda^{1/2} Q_C^{-1} A_C, Q_C^{-1} D_C Q_C^{-1}). \quad \blacksquare$$

A_C, D_C, and Q_C can be estimated by using methods like those described in Section 3.3.3. Let $h_n^* = O(n^{-\delta/(2s+1)})$ for some δ satisfying $0 < \delta < 1$. Then A_C is estimated consistently by

$$A_{nC} = \left(\frac{1}{h_n^*}\right)^s \left[\frac{\pi_1}{n_1} \sum_{\substack{i=1 \\ Y_i=1}}^{n} \frac{\tilde{X}_i}{h_n^*} K'\left(\frac{X_i b_n}{h_n^*}\right) - \frac{\pi_0}{n_0} \sum_{\substack{i=1 \\ Y_i=0}}^{n} \frac{\tilde{X}_i}{h_n^*} K'\left(\frac{X_i b_n}{h_n^*}\right)\right].$$

D_C, and Q_C are estimated consistently by

$$D_{nC} = \left(\frac{\pi_1}{q_1} + \frac{\pi_1}{q_1}\right)\left[\frac{\pi_1}{n_1 h_n} \sum_{\substack{i=1 \\ Y_i=1}}^{n} \tilde{X}_i{}' \tilde{X}_i K'\left(\frac{\tilde{X}_i b_n}{h_n}\right)\right.$$

$$\left. + \frac{\pi_0}{n_0 h_n} \sum_{\substack{i=1 \\ Y_i=0}}^{n} \tilde{X}_i{}' \tilde{X}_i K'\left(\frac{\tilde{X}_i b_n}{h_n}\right)\right],$$

and

$$Q_{nC} = \frac{\partial^2 S_{sms,CB}(b_n)}{\partial b \partial b'}.$$

The formula for D_{nC} corrects equation (6.35) of Horowitz (1993b), which omits the factor $(\pi_1/q_1 + \pi_0/q_0)$.

As in smoothed maximum score estimation with random samples, an asymptotically unbiased smoothed maximum score estimator of β for choice-based samples is given by

$$\hat{b}_n = \tilde{b}_n + \left(\frac{\lambda}{n}\right)^{s/(2s+1)} Q_{nC}^{-1} A_{nC}.$$

The asymptotically optimal value of λ is given by (3.31) but with A, D, and Q replaced with A_C, D_C, and Q_C.

Tests of hypotheses about β can be carried out by replacing A_n, D_n, and Q_n with A_{nC}, D_{nC}, and Q_{nC} in the formulae for the random-sample t and χ^2 statistics. Bootstrap versions of the test statistics can be obtained by sampling the estimation data conditional on Y. That is, one draws of n_1 observations randomly with replacement from the subset of the estimation data for which $Y = 1$, and one draws n_0 observations randomly with replacement from the subset of the estimation data for which $Y = 0$.

In a choice-based sample, the analyst chooses n_1 and n_0 or, equivalently, q_1 and q_0. It is useful, therefore, to ask how this can be done so as to minimize the asymptotic mean-square error or variance of the resulting estimator. This question can be answered for the smoothed maximum score estimator, whose asymptotic variance is known. To find the answer, observe that the right-hand side of (3.36) depends on q_1 and q_0 only through the factor $(\pi_1/q_1 + \pi_0/q_0)$ that multiplies D_C. Therefore, the asymptotic mean-square estimation errors and asymptotic variances of all components $\tilde{b}_n$ are minimized by choosing q_1 and q_0 to minimize $(\pi_1/q_1 + \pi_0/q_0)$.

3.4.2 Panel Data

Panel data consist of observations on individuals at each of several discrete points in time. Thus, there are two or more observations of each individual. The observations of the same individual at different times may be correlated, even if individuals are sampled independently. For example, there may be unobserved attributes of individuals that affect choice and remain constant over time. This section shows how to carry out maximum score and smoothed maximum score estimation of the following binary-response model for panel data:

$$(3.37a) \qquad Y_{it} = \begin{cases} 1 & \text{if } Y_{it}^* > 0 \\ 0 & \text{otherwise} \end{cases}$$

where

$$(3.37b) \qquad Y_{it}^* = X_{it}\beta + U_i + \varepsilon_{it}.$$

In this model, Y_{it} is the dependent variable corresponding to individual i at time t ($i = 1, ..., n$; $t = 1, ..., T$), X_{it} is a vector of explanatory variables corresponding to individual i at time t, β is a conformable vector of constant parameters, U_i is an unobserved random variable that is constant over time, and ε_{it} is an unobserved random variable that varies across both individuals and time. The random variable U_i represents unobserved, time-invariant attributes of individual i, whereas the random variable ε_{it} represents unobserved variables influencing choice that vary across both individuals and time. It is typical in panel data to have observations of many individuals but at only a few points in time. Accordingly, asymptotics will be carried out by assuming that $n \to \infty$ but that T is fixed.

Heckman (1981a, 1981b) discusses estimation of β in parametric models. Manski (1987) has shown how the maximum score method can be used to estimate β without assuming that the distributions of U and ε belong to known, finite-dimensional parametric families. Indeed, the maximum score method places no restrictions on the distribution of U and permits arbitrary serial dependence among the random variables $\{\varepsilon_{it}: t = 1, ..., T\}$ for each i.

The estimator is most easily explained by assuming that $T = 2$. The generalization to larger values of T is straightforward. Assume that for any i, ε_{i1} and ε_{i2} have identical distributions conditional on (X_{it}, U_i; $t = 1, 2$). Define $W_i = X_{i2} - X_{i1}$ and $\eta_i = \varepsilon_{i2} - \varepsilon_{i1}$. Then it follows from (3.37b) that

$$(3.38a) \qquad Y_{i2}^* - Y_{i1}^* = W_i\beta + \eta_i .$$

Moreover, median($\eta_i | W_i = w_i$) = 0 because ε_{i2} and ε_{i1} have the same distributions conditional on (X_{i2}, X_{i1}). Now define

$$(3.38b) \qquad \widetilde{Y}_i = \begin{cases} 1 & \text{if } Y_{i2}^* - Y_{i1}^* > 0 \\ 0 & \text{otherwise} \end{cases}$$

By comparing (3.38) with (3.1), it can be seen that β in the panel-data model could be estimated by the maximum score method if $\widetilde{Y}_i$ were observable. But

$$Y_{i2} - Y_{i1} = 2\widetilde{Y}_i - 1$$

whenever $Y_{i2} \neq Y_{i1}$, so $\widetilde{Y}_i$ is observable if $Y_{i2} \neq Y_{i1}$. Consider, therefore, the estimator of β that is obtained by solving the problem

$$(3.39) \qquad \underset{\substack{b_1 = 1 \\ b \in B}}{\text{maximize:}} \; S_{ms,pan}(b) \equiv n^{-1} \sum_{i=1}^{n} (Y_{i2} - Y_{i1}) I(W_i b \geq 0) .$$

The resulting estimator, b_n, is simply the maximum score estimator of β based on observations for which $Y_{i2} \neq Y_{i1}$. Therefore, arguments similar to those leading to Theorem 3.4 can be used to obtain conditions under which b_n is consistent for β. The result is given in the following theorem, which is proved in Manski (1987). Define $k = \dim(W_i)$ for each $i = 1, ..., n$.

Theorem 3.9: Let model (3.37) hold. Let B be a compact subset of $\Re^k$. Let the first component of β equal one. Let b_n solve problem (3.39). Then $b_n \rightarrow \beta$ almost surely as $n \rightarrow \infty$ if the following conditions hold:

(a) For each i, the random variables ε_{i2} and ε_{i1} are identically distributed conditional on (X_{i2}, X_{i1}, U_i). The support of the common distribution is $(-\infty,\infty)$ for all (X_{i2}, X_{i1}, U_i).

(b) The support of the distribution of W_i $(i = 1, ..., n)$ is not contained in any proper linear subspace of $\Re^k$.

(c) For almost every $\tilde{w} = (w_2, ..., w_k)$, the distribution of W_1 conditional on $(W_2, ..., W_k) = \tilde{w}$ has everywhere positive density.

(d) The random vector $\{Y_{it}, X_{it}: t = 1, 2\}$ has the same distribution for each $i = 1, ..., n$. The data consist of an independent random sample from this distribution. ■

The second sentence of condition (a) in Theorem 3.9 insures that the event $Y_{i2} \neq Y_{i1}$ has nonzero probability. The other conditions are similar to the conditions of Theorem 3.4 and are needed for reasons explained in Section 3.3.2.

Observe that Theorem 3.9 places no restrictions on the distribution of U_i. In particular, U_i may be correlated with X_{i2} and X_{i1}. In this respect, the maximum score estimator for panel data is analogous to the so-called fixed-effects estimator for a linear, mean-regression model for panel data. As in the fixed-effects model, a component of β is identified only if the corresponding component of $X_{i2} - X_{i1} \neq 0$ with nonzero probability. Thus, any intercept component of β is not identified. Similarly, components of β corresponding to explanatory variables that are constant over time within individuals are not identified. As in the linear model, the effects of explanatory variables that are constant over time cannot be separated from the effects of the individual effect U_i under the assumptions of Theorem 3.9.

To extend the maximum score estimator to panels with $T > 2$ define $W_{itr} = X_{it} - X_{ir}$ $(t, r = 2, ..., T)$. Then β can be estimated consistently by solving

$$\underset{\substack{b_1=1 \\ b\in B}}{\text{maximize:}}\ n^{-1}\sum_{i=1}^{n}\sum_{t=2}^{T}\sum_{r<t}(Y_{it}-Y_{ir})I(W_{itr}b\geq 0)\,.$$

See Charlier, *et al.* (1995).

The panel-data maximum score estimator, like the forms of the maximum score estimator that are discussed in Sections 3.3.3 and 3.4.1, has a discontinuous objective function that can be smoothed. The smoothed estimator solves

$$\text{(3.39)}\qquad \underset{\substack{b_1=1 \\ b\in B}}{\text{maximize:}}\ S_{sms,PD}(b)\equiv n^{-1}\sum_{i=1}^{n}\sum_{t=2}^{T}\sum_{r<t}(Y_{it}-Y_{ir})K\left(\frac{W_{itr}b}{h_n}\right),$$

where K is the integral of a kernel function and $\{h_n\}$ is a sequence of bandwidths. To analyze this estimator, let $\tilde{\beta}$ denote the vector consisting of all components of β except the first. Let $\tilde{b}_n$ denote the resulting estimator of $\tilde{\beta}$. Charlier (1994) and Charlier, *et al.* (1995) use Taylor series methods like those used to prove Theorem 3.5 to show that $\tilde{b}_n$ is asymptotically normally distributed after centering and normalization. To state the result, define $W_{itr}^{(1)}$ to be the first component of W_{itr} and $\tilde{W}_{itr}$ to be the vector consisting of all components of W_{itr} except the first. Let $s\geq 2$ be an integer. Also define $Z_{tr}=W_{\bullet tr}\beta$,

$$A_{PD}=$$

$$-2\alpha_A\sum_{t=2}^{T}\sum_{r<t}\sum_{j=1}^{s}\frac{1}{j!(s-j)!}E\big[F^{(j)}(0|0,\tilde{W}_{\bullet tr},Y_t\neq Y_r)$$

$$\times\, p^{(s-j)}(0|\tilde{W}_{\bullet tr},Y_t\neq Y_r|Y_t\neq Y_r)\big]P(Y_t\neq Y_r),$$

$$D_{PD}=\alpha_D\sum_{t=2}^{T}\sum_{r<t}E\big[\tilde{W}_{\bullet tr}'\tilde{W}_{\bullet tr}\,p(0|\tilde{W}_{\bullet tr},Y_t\neq Y_r)|Y_t\neq Y_r\big]P(Y_t\neq Y_r)\,,$$

and

$$Q_{PD} =$$

$$2\sum_{t=2}^{T} \sum_{r<t} E\left[\widetilde{W}'_{\bullet tr}\widetilde{W}_{\bullet tr} F^{(1)}(0|0, \widetilde{W}_{\bullet tr}, Y_t \neq Y_r) p(0|\widetilde{W}_{\bullet tr}, Y_t \neq Y_r) | Y_t \neq Y_r\right].$$

$$\times P(Y_t \neq Y_r).$$

Make the following assumptions:

Assumption PD1: For each i, the random variables ε_{it} ($t = 1, ..., T$) are identically distributed conditional on ($X_{i1}, ..., X_{iT}, U_i$). The support of the common distribution is $(-\infty,\infty)$ for all ($X_{i1}, ..., X_{iT}, U_i$).

Assumption PD2: For all $t = 1, ..., T$ and $r = 1, ..., t - 1$ the support of W_{itr} ($i = 1, ..., n$) is not contained in any proper linear subspace of $\Re^k$.

Assumption PD3: For all $t = 1, ..., T$; $r = 1, ..., T - 1$; and almost every $\widetilde{w}_{tr}$; the distribution of $W_{itr}^{(1)}$ conditional on $\widetilde{W}_{itr} = \widetilde{w}$ has everywhere positive density.

Assumption PD4: The random vector $\{Y_{it}, X_{it}: t = 1, ..., T\}$ has the same distribution for each $i = 1, ..., n$. The data consist of an independent random sample from this distribution.

Assumption PD5: $\widetilde{\beta} \equiv (\beta_2, ..., \beta_k)'$ is an interior point of a compact subset $\widetilde{B}$ of $\Re^{k-1}$.

Assumption PD6: The components of $\widetilde{W}$ and of the matrices $\widetilde{W}'\widetilde{W}$ and $\widetilde{W}'\widetilde{W}\widetilde{W}'\widetilde{W}$ have finite first absolute moments.

Assumption PD7: (a) K is twice differentiable everywhere, $|K'(\bullet)|$ and $|K''(\bullet)|$ are uniformly bounded, and each of the following integrals over $(-\infty, \infty)$ is finite:

$$\int [K'(v)]^4 dv, \; \int [K''(v)]^2 dv, \; \int |v^2 K''(v)| dv.$$

(b) For some integer $s \geq 2$ and each integer j ($1 \leq j \leq s$)

$$\int_{-\infty}^{\infty} v^j K'(v)dv = 0 \ \text{ if } j < s$$

$$= d \ \text{(nonzero) if } j = s.$$

(c) For each integer j between 0 and s, any $\eta > 0$, and any sequence $\{h_n\}$ converging to 0

$$\lim_{n \to \infty} h_n^{j-s} \int_{|h_n v| > \eta} |v^j K'(v)| dv = 0$$

and

$$\lim_{n \to \infty} h_n^{-1} \int_{|h_n v| > \eta} |K''(v)| dv = 0 .$$

Assumption PD8: For each integer j such that $1 \leq j \leq s - 1$, all z in a neighborhood of 0, almost every $\widetilde{w}_{ts}$, $y_t \neq y_s$, and some $M < \infty$, $p^{(j)}(z_{tr}| \widetilde{w}_{tr}, y_t \neq y_r)$ exists and is a continuous function of z_{tr} satisfying $|p^{(j)}(z_{tr}| \widetilde{w}_{tr}, y_t \neq y_r)| < M$ for all z_{tr}. In addition $|p(z_{tr}, z_{km}| \widetilde{w}_{tr}, \widetilde{w}_{km}, y_t \neq y_r, y_k \neq y_m)| < M$ for all (z_{tr}, z_{km}) and almost every $(\widetilde{w}_{tr}, \widetilde{w}_{km})$, $y_t \neq y_r$, and $y_k \neq y_m$.

Assumption PD9: For each integer j such that $1 \leq j \leq s$, all z in a neighborhood of 0, almost every $\widetilde{w}_{tr}$, $y_t \neq y_r$, and some $M < \infty$, $F^{(j)}(-z_{tr}|z_{tr}, \widetilde{w}_{tr}, y_t \neq y_r)$ exists and is a continuous function of z_{tr} satisfying $|F^{(j)}(-z_{tr}|z_{tr}, \widetilde{w}_{tr}, y_t \neq y_r)| < M$.

Assumption PD10: The matrix Q_{PD} is negative definite.

These assumptions are analogous to the ones made in Theorem 3.5. The following theorem is proved in Charlier (1994).

Theorem 3.10: Let model (3.37) and assumptions PD1-PD10 hold for some $s \geq 2$. Assume that $nh_n^{2s+1} = \lambda$ for some finite $\lambda > 0$. Let b_n solve problem (3.38). Then $b_n \to \beta$ almost surely as $n \to \infty$, and

$$(nh_n)^{1/2}(\widetilde{b}_n - \widetilde{\beta}) \to^d N(-\lambda^{1/2} Q_{PD}^{-1} A_{PD}, Q_{PD}^{-1} D_{PD} Q_{PD}^{-1}) .$$

Let Ω be a nonstochastic, positive-semidefinite matrix such that $A_{PD}Q_{PD}^{-1}\Omega Q_{PD}^{-1}A_{PD} \neq 0$. Let E_A denote the expectation with respect to the asymptotic distribution of $(nh_n)^{1/2}(\tilde{b}_n - \tilde{\beta})$, and $MSE \equiv E_A(\tilde{b}_n - \tilde{\beta})'\Omega\tilde{b}_n - \tilde{\beta})$. MSE is minimized by setting $\lambda = \lambda^*$, where

$$\lambda^* = \frac{\text{trace }(Q_{PD}^{-1}\Omega Q_{PD}^{-1}D_{PD})}{2sA_{PD}'Q_{PD}^{-1}\Omega Q_{PD}^{-1}A_{PD}}. \quad \blacksquare$$

The quantities A_{PD}, D_{PD}, and Q_{PD} can be estimated consistently as follows. Define

$$a_{its}(b,h) = \frac{\tilde{W}_{its}}{h_n} K'\left(\frac{W_{its}b}{h}\right)(Y_{it} - Y_{is}).$$

Let $\{h_n^*\}$ be a sequence of positive numbers satisfying $h_n^* \propto n^{-\delta/(2s+1)}$, where $0 < \delta < 1$. Then A_{PD}, D_{PD}, and Q_{PD}, respectively, are estimated consistently by

$$A_{n,PD} = \frac{1}{n(h_n^*)^{s+1}} \sum_{i=1}^{n} \sum_{t=2}^{T} \sum_{s<t} \tilde{W}_{its}' K'\left(\frac{X_i b_n}{h_n^*}\right)(Y_{it} - Y_{is}),$$

$$D_{n,PD} = \frac{h_n}{n} \sum_{i=1}^{n} \sum_{t=2}^{T} \sum_{s<t} a_{its}(b_n, h_n)' a_{its}(b_n, h_n),$$

and

$$Q_{n,PD} = \frac{\partial^2 S_{sms,PD}(b_n)}{\partial \tilde{b}\, \partial \tilde{b}'}.$$

As in smoothed maximum score estimation based on cross-sectional data, an asymptotically unbiased smoothed maximum score estimator of β for panel data is given by

$$\hat{b}_n = \tilde{b}_n + \left(\frac{\lambda}{n}\right)^{s/(2s+1)} Q_{n,PD}^{-1} A_{n,PD}.$$

The asymptotically optimal value of λ is given by (3.31) but with A, D, and Q replaced with A_{PD}, D_{PD}, and Q_{PD}. Tests of hypotheses about β can be carried out by replacing A_n, D_n, and Q_n with $A_{n,PD}$, $D_{n,PD}$, and $Q_{n,PD}$ in the formulae for the random-sample t and χ^2 statistics.

3.4.3 Ordered-Response Models

An ordered-response model is a modification of the binary-response model in which the dependent variable Y takes more than two discrete values. The most frequently used form of this model is

$$(3.41a) \qquad Y = I(\alpha_{m-1} < Y^* \le \alpha_m); \quad m = 1, \ldots, M$$

where

$$(3.41b) \qquad Y^* = X\beta + U ,$$

α_m (m = 0, ..., M) are constants, $-\infty = \alpha_0 < \alpha_1 < \ldots < \alpha_{M-1} < \alpha_M = \infty$, Y is the dependent variable, X is a vector of explanatory variables, β is a conformable vector of constant parameters, and Y^* and U are unobserved random variables. In general, the constants α_m may be known or unknown to the analyst.

Assume for the moment that the α_m are known. Consider the problem of estimating β from a random sample of (Y, X). If the distribution of U is known up to a finite-dimensional parameter, then β can be estimated by maximum likelihood. See, for example, Davidson and MacKinnon (1993). Kooreman and Melenberg (1989) and Lee (1992) have developed versions of the maximum score estimator that permit estimation of β without assuming that the distribution of U belongs to a known, finite-dimensional parametric family.

To describe the semiparametric estimator, assume that median($U|x$) = 0 for almost every x. Define the random variable W by

$$W = \sum_{m=1}^{M} I(Y^* > \alpha_m) .$$

Note that W is observed by virtue of (3.41a). In addition, median($U|x$) = 0 implies that median($Y^*|x$) = $x\beta$. Therefore, it follows from (3.41a) that

$$(3.42) \qquad \text{median}\,(W|\,x) = \sum_{m=0}^{M} I(x\beta > \alpha_m) .$$

Let the data, $\{Y_i, X_i: i = 1, \ldots, n\}$ be a random sample of (Y, X). Consider, the modified maximum score estimator of β that is obtained by solving the problem

$$(3.43) \qquad \underset{b \in B}{\text{maximize:}}\; S_{n,OR}(b) \equiv \frac{1}{n}\sum_{i=1}^{n} |W_i - \sum_{m=0}^{M} I(X_i b > \alpha_m)| ,$$

where B is the parameter set. Let b_n be the resulting estimator. Observe that because of (3.42), b_n amounts to a median-regression estimator of β. Consistency of this estimator is established in the following theorem, which is proved in Kooreman and Melenberg (1989).

Theorem 3.11: Let model (3.41) hold with median$(U|x) = 0$ for all x in the support of X.. Assume that B is a compact subset of $\Re^k$. Let b_n solve problem (3.43). Then $b_n \to \beta$ almost surely as $n \to \infty$ if

(a) $\{Y_i, X_i\}$ is a random sample of size n from the distribution of (Y, X).

(b) The support of the distribution of X is not contained in any proper linear subspace of $\Re^k$.

(c) The first component of β is nonzero. For almost every $\tilde{x} = (x_2, \ldots, x_k)$ the distribution of X_1 conditional on $\tilde{X} = \tilde{x}$ has an everywhere positive density.

(d) The bounds $\{\alpha_m: m = 0, ..., M\}$ are known constants, and $M > 2$. ■

These assumptions are similar to those of Theorem 3.4. The main difference is that the ordered-response model does not require a scale normalization for β. This is because levels of Y^* are observable in the ordered-response model. In contrast, only the sign of Y^* is observable in the binary-response model.

Lee (1992) showed that the maximum score estimator for an ordered-response model can be extended to the case of unknown α_m's by normalizing the absolute value of the first component of β to be 1, normalizing $\alpha_1 = 0$, and minimizing the right-hand side of (3.42) with respect to α_m $(m = 1, ..., M)$ as well as β. Melenberg and van Soest (1996) have derived a smoothed version of the ordered-response maximum score estimator with unknown α_m's . This estimator is obtained by replacing the indicator function $I(X_i b > \alpha_m)$ on the right-hand side of (3.42) with the smooth function $K[(X_i b - a_m)/h_n]$, where $\{h_n\}$ is a sequence of bandwidths. Define $W_{im} = I(Y_i^* > a_m)$. Then the smoothed estimator solves

$$\underset{b \in B,\, a_1 = 0}{\text{maximize:}}\; S_{n,SOR}(b,a) \equiv \frac{1}{n}\sum_{i=1}^{n}\sum_{m=1}^{M}(2W_{im} - 1)K\left(\frac{X_i b - a_m}{h_n}\right).$$

Melenberg and van Soest provide informal arguments showing that the centered, normalized, smoothed maximum score estimator of $\tilde{\beta}$ and the α_m for ordered response models is asymptotically normally distributed under regularity conditions similar to those of Theorem 3.5.

To state the result of Melenberg and van Soest, define $Z_m = X\beta - \alpha_m$. Assume that $\alpha_1 = 0$ to achieve location normalization, and let $\tilde{\alpha}$ be the vector $(\alpha_2, \ldots, \alpha_{M-1})'$. Let $\tilde{a}_n$ be the smoothed maximum score estimator of $\tilde{\alpha}$. Let $\tilde{X}$ denote the vector consisting of all components of X except the first, and let $p_m(\bullet|\tilde{x})$ denote the probability density function of Z_m conditional on $\tilde{X} = \tilde{x}$. Let $F(\bullet|z_r, \tilde{x})$ denote the CDF of U conditional on $Z = z_r$ and $\tilde{X} = \tilde{x}$. Define $\tilde{X}_m = (\tilde{x}, -d_m')'$, where $d_m = 0$ if $m = 1$, $d_m = (0, \ldots, 1, 0, \ldots, 0)'$ if $2 \le m \le M - 1$, and 1 is the $(r - 1)$'th component of the vector. Then define the matrix A_{OR} by

$$A_{OR} = \sum_{m=1}^{M-1} A_m ,$$

where

$$A_m = -2\alpha_A \sum_{j=1}^{s} \left\{ \frac{1}{j!(s-j)!} E\left[F^{(j)}(0|0, \tilde{X}) p_m^{(s-j)}(0|\tilde{X})\tilde{X}_m\right]\right\},$$

and $s \ge 2$ is an integer. Also define

$$D_{OR} = \sum_{m=1}^{M-1} D_m + \sum_{m=1}^{M-1} \sum_{k \ne m} D_{mk} ,$$

where for $m, k = 1, \ldots, M - 1$

$$D_m = \alpha_D E\left[\tilde{X}'\tilde{X} p_m(0|\tilde{X})\right]$$

$$D_{mk} = \alpha_D E\left\{\left[P_{mk}(1|0, \tilde{X}) - P_{mk}(-1|0, \tilde{X})\right]\tilde{X}_k' \tilde{X}_m p_m(0|\tilde{X}) p_k(0|\tilde{X})\right\},$$

$$P_{mk}(1|0, \tilde{x}) = P(Y_{im} = Y_{ik} | z_m = 0, z_s = 0, \tilde{x}),$$

and

$$P_{mk}(-1|0, \tilde{x}) = P(Y_{im} \ne Y_{ik} | z_m = 0, z_s = 0, \tilde{x}).$$

Finally, define

$$Q_{OR} = \sum_{m=1}^{M-1} Q_m ,$$

where

$$Q_m = 2E\left[\widetilde{X}'_m \widetilde{X}_m F^{(1)}(0|0,\widetilde{X}) p_m(0|\widetilde{X})\right].$$

Melenberg and van Soest (1996) provide informal arguments showing that if $nh_n^{2s+1} \to \lambda$ for some $s \geq 2$, then under regularity conditions similar to those of Theorem 3.5, the ordered-response smoothed maximum score estimator satisfies

$$(nh_n)^{1/2}\left[(\widetilde{b}'_n, \widetilde{a}'_n)' - (\widetilde{\beta}', \widetilde{\alpha}')'\right] \to^d N(-\lambda^{1/2} Q_{OR}^{-1} A_{OR}, Q_{OR}^{-1} D_{OR} Q_{OR}^{-1}).$$

A_{OR}, D_{OR}, and Q_{OR} can be estimated consistently by replacing the derivatives of the summand of S_{sms} with the derivatives of the summand of $S_{n.SOR}$ in the formulae for A_n, D_n, and Q_n in smoothed maximum score estimation of a binary-response model.

3.5 An Empirical Example

This section presents an empirical example that illustrates the use of the maximum score and smoothed maximum score estimators for a binary response model. The example is taken from Horowitz (1993a) and consists of estimating a model of the choice between automobile and transit for the trip to work. The data consist of 842 observations of work trips that were sampled randomly from the Washington, D.C., area transportation study. The dependent variable is 1 if automobile was used for the trip and 0 if transit was used. The independent variables are an intercept (*INT*), the number of cars owned by the traveler's household (*CARS*), transit out-of-vehicle travel time minus automobile out-of-vehicle travel time (*DOVTT*), transit in-vehicle travel time minus automobile in-vehicle travel time (*DIVTT*), and transit fare minus automobile travel cost (*DCOST*). *DOVTT* and *DIVTT* are in units of minutes, and *DCOST* is in units of dollars. Scale normalization is achieved by setting $\beta_{DCOST} = 1$.

The model is specified as in (3.1). The coefficients were estimated by parametric maximum likelihood using a binary probit model, the maximum score estimator, and the smoothed maximum score estimator. In smoothed maximum score estimation, K was the standard normal distribution function. The smoothed maximum score estimates are bias corrected.

The estimation results are shown in Table 3.2. The quantities in parentheses are the half-widths of nominal 90% confidence intervals. For the binary probit model, these were obtained using standard asymptotic distribution theory. For the maximum score and smoothed maximum score estimator, the confidence intervals were obtained using the bootstrap methods described in Sections 3.3.2 and 3.3.3. With the smoothed maximum score estimator, the resulting 1 - α confidence interval for the j'th component of $\widetilde{\beta}$ is

$$\widetilde{b}_{nj} - z_\alpha^* V_{nj}^{1/2} \leq \widetilde{\beta}_j \leq \widetilde{b}_{nj} + z_\alpha^* V_{nj}^{1/2},$$

where z_α^* is the (1 - α) quantile of the distribution of the bootstrap t statistic, t^*. Half-widths of confidence intervals, rather than standard errors, are used to indicate estimation precision for two reasons. First, as was discussed in Section 3.3.2, the maximum score estimator is not asymptotically normally distributed. Therefore, there is no simple relation between the standard error of the maximum score estimator and either the width of a confidence interval or the outcome of a hypothesis test. Second, although the smoothed maximum score estimator is asymptotically normally distributed, the bootstrap provides a higher-order approximation to its distribution, and the higher-order approximation is non-normal.

The estimates of β obtained by the three different estimation methods differ by more than a factor of two. Thus, the different methods yield very different results. The maximum score and smoothed maximum score estimates indicate much greater sensitivity of choice to *CARS* and *DIVTT* than do the probit estimates. The half widths of the confidence intervals for non-intercept coefficients are largest for the maximum score estimates and narrowest for the probit estimates. This is consistent with the relative rates of convergence of the three estimators, which are $n^{-1/3}$ for the maximum score estimator, $n^{-2/5}$ for the smoothed maximum score estimator, and $n^{-1/2}$ for the probit model.

Table 3.2: Estimated Parameters of a Work-Trip Mode-Choice Model

Estimator	INT	CARS	DOVTT	DIVTT
Probit	-0.628	1.280	0.034	0.006
	(0.382)	(0.422)	(0.024)	(0.006)
Smooth. Max. Score	-1.576	2.242	0.027	0.014
	(0.766)	(0.749)	(0.031)	(0.009)
Max. Score	-1.647	2.252	0.041	0.011
	(0.118)	(2.106)	(0.049)	(0.018)

Source: Horowitz (1993a). The coefficient of *DCOST* is 1 by scale normalization.

The coefficient of *CARS* is statistically significantly different from 0 at the 0.10 level according to all estimation methods. That is, the 90% confidence intervals for β_{CARS} do not contain 0. However, the smoothed maximum score estimates yield inferences about β_{DOVTT} and β_{DIVTT} that are different from those obtained from the probit model. The smoothed maximum score method gives a smaller point estimate of β_{DOVTT} and a larger point estimate of β_{DIVTT} than does the probit model. In addition, β_{DOVTT} is not significantly different from 0 at the 0.10 level and β_{DIVTT} is significantly different from 0 according to the smoothed maximum score estimates, whereas the opposite is the case according to the probit estimates. The non-significance of the smoothed maximum score estimate of β_{DOVTT} does not necessarily indicate that the true value of β_{OVTT} is close to 0 because the smoothed maximum score estimate is relatively imprecise. However, the probit model is nested in the model assumed by the smoothed maximum score estimator. Therefore, the differences between the probit and smoothed maximum score estimates suggest that the probit model is misspecified and that its estimates are misleading.

In fact the probit model is rejected by a variety of specification tests. These include likelihood-ratio, Wald, and Lagrangian multiplier tests against a random-coefficients probit model. Thus the differences between the probit and smoothed maximum score estimates reflect genuine features of the data-generation process, not just random sampling errors.

Chapter 4
Deconvolution Problems

This chapter is concerned with estimating the distribution of a random variable U when one observes realizations not of U but of $W \equiv U + \varepsilon$, where ε is random variable that is independent of U. Such estimation problems are called *deconvolution problems* because the distribution of the observed random variable, W, is the convolution of the distributions of U and ε. Estimating the distribution of U requires deconvoluting the distribution of the observed random variable W.

Deconvolution problems arise, among other places, in models of measurement error. For example, suppose that W is the measured value of a variable, U is the true value, and ε is the measurement error. If error-free measurement is not possible, then estimating the distribution of the true variable U is a problem in deconvolution. Section 4.1 shows how to solve this problem.

Deconvolution problems also arise in mean-regression models for panel data. For example, consider the model

$$(4.1) \quad Y_{jt} = X_{jt}\beta + U_j + \varepsilon_{jt}, \quad j = 1,\ldots,n;\ t = 1,\ldots,T$$

In this model, Y_{jt} is the observed value of the dependent variable Y for individual j at time t, X_{jt} is the observed value of a vector of explanatory variables X for individual j at time period t, and β is a vector of constant parameters. U_j is the value of an unobserved random variable U that varies across individuals but is constant over time, and ε_{jt} is the value of an unobserved random variable ε that varies across both individuals and time. Suppose one wants to estimate the probability distribution of Y_{js} at time period $s > T$ conditional on observations of Y_{jt} and X_{jt} at time periods $t \leq T$. For example, Y_{js} may be the income of individual j at time s, and one may want to estimate the probability that individual j's future income exceeds a specified value conditional on observations of this individual's past income. To solve this problem, it is necessary to estimate the distributions of U and ε, which is a deconvolution problem. Section 4.2 shows how to carry out deconvolution for the panel-data model (4.1).

4.1 A Model of Measurement Error

This section is concerned with estimating the probability density function of the random variable U in the model

$$(4.2) \qquad W = U + \varepsilon .$$

The data consist of a random sample of observations of W: $\{W_j: j = 1, ..., n\}$. It is assumed that ε is independent of U and that ε has a known probability density function f_ε. In an application, knowledge of f_ε might be obtained from a validation data set. A validation data set contains observations of both U and W, thereby permitting estimation of the probability distribution of ε. Typically, the validation data set is much smaller than the full estimation data set that contains observations of W but not U. Therefore, the validation data set is not a substitute for the full estimation data set.

The procedure described here for estimating the distribution of U has two main steps. The first is to express the density of U as a functional of the distributions. of W and ε. The second is to replace the unknown distribution of W with a suitable sample analog.

To take the first estimation step, define ψ_W, ψ_U, and ψ_ε, respectively, to be the characteristic functions of the distributions of W, U, and ε. See Rao (1973) and Stuart, and Ord (1987) for discussions of characteristic functions and their properties. Assume that $\psi_\varepsilon(\tau) \neq 0$ for all finite, real τ. Then for any finite, real τ, $\psi_W(\tau) = \psi_U(\tau)\psi_\varepsilon(\tau)$ (because U and ε are assumed to be independent), and $\psi_U(\tau) = \psi_W(\tau)/\psi_\varepsilon(\tau)$. Let f_U denote the probability density function of U. It follows from the inversion formula for characteristic functions that

$$(4.3) \qquad f_U(u) = \frac{1}{2\pi} \int_{-\infty}^{\infty} e^{-i\tau u} \frac{\psi_W(\tau)}{\psi_\varepsilon(\tau)} d\tau ,$$

where $i = (-1)^{1/2}$. Equation (4.3) provides the desired expression for f_U in terms of the distributions of W and ε. This completes the first estimation step.

The only unknown quantity on the right-hand side of (4.3) is ψ_W. Therefore, the second estimation step consists of replacing ψ_W with a suitable sample analog. One possibility is the empirical characteristic function of W. This is given by

$$\psi_{nW}(\tau) = \frac{1}{n} \sum_{j=1}^{n} \exp(i\tau W_j) .$$

One could now consider replacing ψ_W with ψ_{nW} in (4.3). In general, however, the resulting integral does not exist. This is because (4.3) assumes that the distribution of W has a density, but the empirical distribution of W is discrete and, therefore, has no density. This problem can be overcome by convoluting the empirical distribution of W with the distribution of a suitable continuously distributed random variable that becomes degenerate as $n \to \infty$. This amounts to kernel smoothing of the empirical distribution of W.

To carry out the smoothing, let ψ_ζ be a bounded characteristic function whose support is [-1,1]. Let ζ be the random variable that has this characteristic function, and let $\{\nu_n\}$ be a sequence of positive constants that converges to 0 as $n \to \infty$. The idea of the smoothing procedure is to use the inversion formula for characteristic functions to estimate the density of the random variable $W + \nu_n\varepsilon$. The resulting smoothed estimator of $f_U(u)$ is

$$(4.4) \qquad f_{nU}(u) = \frac{1}{2\pi} \int_{-\infty}^{\infty} e^{-i\tau u} \frac{\psi_{nW}(\tau)\psi_\zeta(\nu_n\tau)}{\psi_\varepsilon(\tau)} d\tau .$$

The integral in (4.4) exists because the integrand is 0 if $|\tau| > 1/\nu_n$. If ν_n does not converge to 0 too rapidly as $n \to \infty$, then $f_{nU}(u)$ is consistent for $f_U(u)$. This is analogous to the situation in kernel nonparametric density estimation, where consistency is achieved if the bandwidth converges to 0 at a rate that is not too fast.

The CDF of U, F_U, can be estimated by integrating f_{nU}. The estimator is

$$F_{nU}(u) = \int_{-M_n}^{u} f_{nU}(u)du ,$$

where $M_n \to \infty$ as $n \to \infty$.

4.1.1 Rate of Convergence of the Density Estimator

The large-sample properties of f_{nU} have been investigated by Carroll and Hall (1988), Fan (1991ab), and Stefanski and Carroll (1990, 1991), among others. The main results concern the rate of convergence of f_{nU} to f_U, which is often very slow. Slow convergence is unavoidable in deconvolution problems and is not an indication that f_{nU} in (4.4) is unsatisfactory. This section states the formal results on the rates of convergence of deconvolution estimators. Section 4.1.2 provides a heuristic explanation of these results..

To state the results, make the following assumptions:

Assumption 1: ψ_ζ is real-valued, has support [-1,1], is symmetrical about 0, and has $m + 2$ bounded, integrable derivatives for some $m > 0$.

Assumption 2: $\psi_\zeta(\tau) = 1 + O(\tau^m)$ as $\tau \to 0$.

Assumption 3: $\psi_\varepsilon(\tau) \neq 0$ for all finite, real τ.

Assumption 4: f_U has m bounded derivatives.

Assumptions 1, 2, and 4 are smoothness and symmetry conditions. Assumptions 1 and 2 can always be satisfied by making a suitable choice of ψ_ζ. Assumption 3 insures that the denominator of the integrand in (4.3) is non-zero.

The following theorems on the rate of convergence of f_{nU} are proved in Fan (1991a). Fan (1991a) also investigates the rates of convergence of estimators of derivatives of f_U, but estimation of derivatives will not be discussed here.

Theorem 4.1: Let assumptions 1-4 hold. Also assume that

$$(4.5) \qquad |\psi_\varepsilon(\tau)||\tau|^{-\beta_0} \exp(|\tau|^\beta/\gamma) \geq d_0$$

as $\tau \to \infty$ for positive constants β_0, β, γ, and d_0. Let $\nu_n = (4/\gamma)^{1/\beta}(\log n)^{-1/\beta}$. Then for any u,

$$E[f_{nU}(U) - f_U(U)]^2 = O[(\log n)^{-2m/\beta}]$$

as $n \to \infty$. Moreover, if

$$(4.6) \qquad |\psi_\varepsilon(\tau)||\tau|^{-\beta_1} \exp(|\tau|^\beta/\gamma) \leq d_1$$

as $\tau \to \infty$ for positive constants β_1 and d_1, and if

$$(4.7) \qquad P(|\varepsilon - x| \leq |x|^{\alpha_0}) = O[|x|^{-(a-\alpha_0)}]$$

as $x \to \pm\infty$ for some α_0 satisfying $0 < \alpha_0 < 1$ and $a > 1 + \alpha_0$, then no estimator of $f_U(u)$ can converge faster than $(\log n)^{-m/\beta}$ in the sense that for every estimator $f_{nU}(u)$

$$E[f_{nU}(u) - f_U(u)]^2 > d(\log n)^{-2m/\beta}$$

for some $d > 0$. ■

The technical condition (4.7) holds if the density of ε satisfies $f_\varepsilon(x) = O(|x|^{-a})$ as $|x| \to \infty$ for some $a > 1$. Thus, (4.7) is a restriction on the thickness of the tails of f_ε.

Theorem 4.2: Let assumptions 1-4 hold. Also assume that

(4.8) $|\psi_\varepsilon(\tau)||\tau|^\beta \geq d_0$

as $\tau \rightarrow \infty$ for positive constants β and d_0. Let $\nu_n = dn^{-1/2[(m+\beta)+1]}$ for some $d > 0$. Then for any u,

$$E[f_{nU}(U) - f_U(U)]^2 = O\{n^{-2m/[2(m+\beta)+1]}\}$$

as $n \rightarrow \infty$. Moreover, if $\psi_\varepsilon^{(j)}$ denotes the j'th derivative of ψ_ε and

$$\left|\psi_\varepsilon^j(\tau)\tau^{-(\beta+j)}\right| \leq d_j; \quad j = 0,1,2$$

as $\tau \rightarrow \infty$ for positive constants β and d_j (j = 0, 1, 2), then no estimator of $f_U(u)$ can converge faster than $n^{-m/(2m+2\beta+1)}$ in the sense that for every estimator $f_{nU}(u)$

$$E[f_{nU}(u) - f_U(u)]^2 > dn^{-2m/(2m+2\beta+1)}$$

for some $d > 0$. ■

Similar results can be obtained regarding the rate of uniform convergence of f_{nU} and the rate of convergence of the integrated mean-square error of f_{nU}. In particular, under the assumptions of Theorem 4.1, $\sup_u|f_{nU}(u) - f_U(u)|$ converges in probability to 0 at the rate given by the theorem. Fan (1991a) also shows that F_{nU} is a consistent estimator of F_U if $M_n = O(n^{1/3})$ as $n \rightarrow \infty$. Moreover, under the assumptions of Theorem 4.1, $F_{nU}(u)$ converges to $F_U(u)$ at the same rate at which $f_{nU}(u)$ converges to $f_U(u)$.

Theorems 4.1 and 4.2 imply that f_{nU} converges to f_U at the fastest possible rate, but that rate can be excruciatingly slow. Under the assumptions of Theorem 4.1, the fastest possible rate of convergence of f_{nU} is a power of $(\log n)^{-1}$. By contrast, parametric estimators usually converge at the rate $n^{-1/2}$. Nonparametric mean-regression estimators converge at the rate $n^{-2s/(2s+d)}$, where d is the dimension of the explanatory variable and s is the number of times that the conditional mean function and the density of the explanatory variables are differentiable.

Theorems 4.1 and 4.2 also imply that the rate of convergence of f_{nU} is controlled mainly by the thickness of the tail of the characteristic function of ε.. Conditions (4.5) and (4.6) of Theorem 4.1 are satisfied by distributions whose characteristic functions have tails that decrease exponentially fast. These include the normal, Cauchy, and Type 1 extreme value distributions. The fastest possible rate of convergence of f_{nU} is logarithmic when ε has one of these distributions. Theorem 4.2 shows that faster rates of convergence of f_{nU}

are possible when the tail of the characteristic function of ε decreases only geometrically fast (e.g., a negative power of its argument), as is assumed in condition (4.8). The Laplace and symmetrical gamma distributions have this property.

4.1.2 Why Deconvolution Estimators Converge Slowly

This section provides a heuristic explanation of why the tail of ψ_ε has such a strong influence on the rate of convergence of f_{nU} and why the rate of convergence can be very slow. To begin, write $f_{nU}(u) - f_U(u)$ in the form

$$f_{nU}(u) - f_U(u) = I_{n1}(u) + I_{n2}(u),$$

where

$$I_{n1}(u) = \frac{1}{2\pi}\int_{-\infty}^{\infty} e^{-i\tau u} \frac{[\psi_{nW}(\tau) - \psi_W(\tau)]\psi_\zeta(\nu_{nU}\tau)}{\psi_\varepsilon(\tau)} d\tau$$

and

$$I_{n2}(u) = \frac{1}{2\pi}\int_{-\infty}^{\infty} e^{-i\tau u} \frac{\psi_W(\tau)\psi_\zeta(\nu_{nU}\tau) - \psi_W(\tau)}{\psi_\varepsilon(\tau)} d\tau$$

$$= \frac{1}{2\pi}\int_{-\infty}^{\infty} e^{-i\tau u} \psi_U(\tau)[\psi_\zeta(\nu_{nU}\tau) - 1] d\tau$$

$$= \frac{1}{2\pi}\int_{-\infty}^{\infty} e^{-i\tau u} \psi_U(\tau)\psi_\zeta(\nu_{nU}\tau) d\tau - f_U(u).$$

$I_{n1}(u)$ is a random variable that captures the effects of random sampling error in estimating ψ_W. $I_{n2}(u)$ is a nonstochastic bias arising from the smoothing of the empirical distribution of W. It is analogous to the bias caused by smoothing in nonparametric density estimation and mean regression.

Now, $\psi_{nW}(\tau) - \psi_W(\tau) = O_p(n^{-1/2})$. Moreover, as $\nu_n \to 0$, the denominator of the integral on the right-hand side of I_{n1} will be smallest when $\tau = \pm 1/\nu_n$. Therefore, the integral can be expected to converge as $n \to \infty$ only if

$$\text{(4.9)} \qquad \frac{1}{n^{1/2}\psi_\varepsilon(1/|\nu_n|)} \to 0$$

as $n \to \infty$. Note that (4.9) is only a necessary condition for convergence of the integral. It is not sufficient in general because the relation $\psi_{nW}(\tau) - \psi_W(\tau) = O_p(n^{-1/2})$ does not hold uniformly over τ. The exact rate of convergence of $I_{n1}(u)$ can be obtained by calculating its variance and applying Chebyshev's inequality. This more elaborate calculation is not needed, however, for the heuristic argument that is made here.

Now consider $I_{n2}(u)$. Observe that $\psi_U(\tau)\psi_\zeta(\nu_n\tau)$ is the characteristic function evaluated at τ of the random variable $U + \nu_n\zeta$. Therefore,

$$(4.10) \qquad I_{n2}(u) = \int f_U(u - \nu_n z) f_\zeta(z)dz - f_U(u),$$

where f_ζ is the probability density function of ζ. By assumption 1, ψ_ζ is real, which implies that f_ζ is symmetrical about 0 and $E(\zeta) = 0$. Therefore, a Taylor series expansion of the integrand of (4.10) gives

$$(4.11) \qquad I_{n2}(u) = -\frac{1}{2}\nu_n^2 f_U''(u)\sigma_\zeta^2 + o(\nu_n^2).$$

Suppose that ε is normally distributed with mean 0. Then $\psi_\varepsilon(\tau) \propto \exp(-a\tau^2)$ for some $a > 0$. Moreover, (4.9) implies that $n^{-1/2}\exp(a/\nu_n^2) \to 0$ or $a/\nu_n^2 - (1/2)(\log n) \to -\infty$ as $n \to \infty$. This relation can hold only if ν_n converges to 0 no faster than $(\log n)^{-1/2}$. But $I_{n2}(u) = O(\nu_n^2)$, so $I_{n2}(u)$ converges to 0 most rapidly when the rate of convergence of ν_n is the fastest possible. Therefore, the fastest possible rate of convergence of $I_{n2}(u)$ is $(\log n)^{-1}$. Since $f_{nU}(u)$ can converge to $f_U(u)$ no faster than $I_{n2}(u)$, the fastest possible rate of convergence of $f_{nU}(u)$ is $(\log n)^{-1}$.

This heuristic result can be compared with the conclusions of Theorem 4.1. In the notation of that theorem, $\beta = 1$. If it is assumed that the unknown f_U belongs to a class of twice differentiable densities ($m = 2$), then the fastest possible rate of convergence of $f_{nU}(u)$ according to Theorem 4.1 is $(\log n)^{-1}$, and this occurs when $\nu_n \propto (\log n)^{-1/2}$. Thus, the heuristic analysis of the previous paragraph is consistent with the conclusions of the theorem.

Now suppose that the tails of ψ_ε converge to 0 at a geometric rate so that $\psi_\varepsilon(\tau) \propto |\tau|^{-\beta}$ for some $\beta > 0$. Then (4.9) implies that $n^{-1/2}\nu_n^{-\beta} \to 0$ as $n \to \infty$, which permits a geometric rate of convergence of ν_n and, therefore, of $f_{nU}(u)$. This result is consistent with the conclusions of Theorem 4.2.

The rate of convergence $n^{-m/(2m + \beta) + 1}$ that is given by Theorem 4.2 is obtained by equating the rates of convergence of the variance and squared bias of $f_{nU}(u)$. An argument that is more precise than the heuristic one made here shows that $\text{Var}[I_{n1}(u)] = O[\nu_n^{-(2\beta + 1)}]$ when $m = 2$. This is larger than the left-hand side of (4.9), which is not surprising because, as has already been explained, (4.9) is only necessary for convergence of $I_{n2}(u)$ to 0, not sufficient.

Setting $O[\nu_n^{-(2\beta+1)}] = O(\nu_n^4)$ yields $\nu_n = O[n^{-1/(2\beta+5)}]$. With this rate of convergence of ν_n, $\mathrm{Var}[I_{n1}(u)]$ and $I_{n2}(u)^2$ are both $O[n^{-4/(2\beta+5)}]$, which is the rate given by Theorem 4.2 for $m = 2$.

When $m > 2$, convergence of $f_{nU}(u)$ can be accelerated by replacing ψ_ζ with the Fourier transform of a higher-order kernel. This causes f_ζ on the right-hand side of (4.11) to be replaced by the higher-order kernel. Taylor series arguments similar to those made in nonparametric density estimation then show that $I_{n2}(u) = O(\nu_n^m)$. As in nonparametric density estimation, existence of higher-order derivatives and use of a higher-order kernel accelerates convergence of $f_{nU}(u)$ by accelerating the rate of convergence of the bias for any given bandwidth sequence.

4.1.3 Asymptotic Normality of the Density Estimator

Asymptotic normality of the standardized form of $f_{nU}(u)$ can be proved by showing that it satisfies the conditions of a triangular-array central limit theorem. See Serfling (1980) for a discussion of triangular-array central limit theorems. Fan (1991b) has proved that if $\nu_n = o[n^{-1/(2m+2\beta+1)}]$, then under the assumptions of Theorem 4.2,

$$(4.12)\qquad \frac{f_{nU}(u) - f_U(u)}{\{Var[f_{nU}(u)]\}^{1/2}} \to^d N(0,1).$$

Obtaining asymptotic normality when the tails of ψ_ε decrease exponentially fast requires strengthening the conditions of Theorem 4.1. The required conditions are stated in the following theorem.

Theorem 4.3: Let assumptions 1-4 hold. Also assume that

(a) There are positive constants β_0, β, γ, d_0 and d_1 such that

$$d_1 \geq |\psi_\varepsilon(\tau)||\tau|^{-\beta_0} \exp(|\tau|^\beta/\gamma) \geq d_0$$

(b) As $|\tau| \to \infty$, either $\mathrm{Re}[\psi_\varepsilon(\tau)] = o\{\mathrm{Im}[\psi_\varepsilon(\tau)]\}$ or $\mathrm{Im}[\psi_\varepsilon(\tau)] = o\{\mathrm{Re}[\psi_\varepsilon(\tau)]\}$.

(c) $\psi_\zeta(\tau)$ has $m + 2$ continuous derivatives. Moreover, $\psi_\zeta(\tau) > d_3(1 - \tau)^{m+3}$ for $\tau \in [1 - \delta, 1)$ and some $d_3 > 0$ and $\delta > 0$.

Let $\nu_n \propto (\log n)^{-1/\beta}$. Then (4.12) holds. ■

To use (4.12) for inference, it is necessary to have an estimator of $\mathrm{Var}[f_{nU}(u)]$. To this end, for any real w, define

$$g_n(w) = \frac{1}{2\pi} \int e^{-i\tau w} \frac{\psi_\zeta(\tau)}{\psi_\varepsilon(\tau / \nu_n)} d\tau$$

and

$$Z_{nj} = \frac{1}{\nu_n} g_n\left(\frac{u - W_j}{\nu_n}\right)$$

Fan (1991b) shows that Var$[n^{1/2} f_{nU}(u)]$ is estimated consistently by

$$(4.13) \quad s_n^2 = \frac{1}{n} \sum_{j=1}^{n} Z_{nj}^2$$

in the sense that $\{\text{Var}[n^{1/2} f_{nU}(u)]\} \to^p 1$. Let s_n be the positive square root of s_n^2. Then (4.12) and (4.13) may be combined to give

$$(4.14) \quad \frac{n^{1/2}[f_{nU}(u) - f_U(u)]}{s_n} \to^d N(0,1)\,.$$

Equation (4.14) can be used in the usual way to form confidence intervals for $f_U(u)$ and test hypotheses about $f_U(u)$. For example, an asymptotic 100(1 - α) percent confidence interval for $f_U(u)$ is

$$(4.15) \quad f_{nU}(u) - z_{\alpha/2} \frac{s_n}{n^{1/2}} \le f_U(u) \le f_{nU}(u) + z_{\alpha/2} \frac{s_n}{n^{1/2}}\,.$$

Because of the slow rates of convergence of deconvolution density estimators, the true coverage probability of the confidence interval (4.15) may be very different from the nominal coverage probability of (1 - α) unless n is very large. Therefore, (4.15) must be used with much caution.

4.1.4 A Monte Carlo Experiment

Because the rate of convergence of f_{nU} under the assumptions of Theorem 4.1 is very slow, it useful to investigate whether f_{nU} can provide useful information about f_U under these assumptions and with samples of the sizes likely to be encountered in applications. Such an investigation can be carried out through Monte Carlo experimentation. This section reports the results of one such experiment.

Data for the experiment were generated by simulation from model (4.2). The sample size is $n = 1000$. The distributions of U and ε are both N(0,1), so Theorem 4.1 applies. If it is assumed that $m = 2$, then the fastest possible rate of convergence in probability of f_{nU} is $(\log n)^{-1}$. The smoothing function ψ_ζ is the fourfold convolution of the uniform density with itself. This is the characteristic function of the density $c[(\sin x)/x]^4$, where c is a normalization constant. The smoothing parameter ν_n was chosen by Monte Carlo to approximately minimize the integrated mean-square error of f_{nU}. Of course, this method is not available in applications. It is possible, however, to construct a resampling method that is available in applications and that mimics the Monte Carlo procedure. The method consists of generating samples of size $m < n$ by sampling the estimation data randomly *without* replacement. Each data set created this way is a random sample from the true population distribution of W. By treating f_{nU} based on a preliminary value of ν_n as if it were the true f_U, one can find the value of ν_m that minimizes the integrated mean-square error of f_{mU} as an estimator of f_{nU}. The resulting ν_m can be rescaled to apply to a sample of size n by setting $\nu_n = [r(m)/r(n)]\,\nu_m$, where $r(n)$ is the rate of convergence of ν_n given by Theorem 4.1 or 4.2.

There were 100 Monte Carlo replications in the experiment. Each replication consisted of computing $f_{nU}(u)$ at 25 different values of u. The Monte Carlo estimates of f_{nU} are summarized in Figure 4.1. The left-hand panel of the figure shows the true f_U (dashed line) and the average of 100 estimates f_{nU} (solid line). The right-hand panel shows the true f_U (dashed line) and 10 individual realizations of f_{nU} (solid lines). It can be seen that although the estimates are flatter than the true f_U, their shapes are qualitatively similar to the shape of the true f_U on average and, in most cases, individually. Thus, at least for the distributions used in this experiment, f_{nU} is a useful estimator of f_U despite its slow rate of convergence.

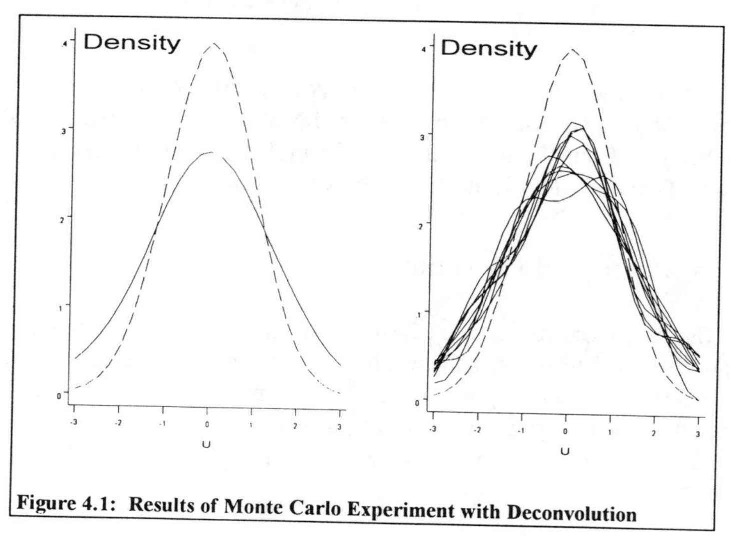

Figure 4.1: Results of Monte Carlo Experiment with Deconvolution

4.2 Models for Panel Data

This section is concerned with estimating the probability density functions f_U and f_ε of the random variables U and ε in model (4.1). Although f_U and f_ε by themselves are rarely of interest in applications, knowledge of these densities is needed to compute certain quantities that are of interest, such as transition probabilities and first passage times.

To illustrate, consider first passage times. The first passage time for individual j in model (4.1) is the smallest t for which Y_{jt} exceeds a specified threshold, say y^*. Suppose one is interested in the first passage time conditional on the initial value of Y for individual j, Y_{j1}, and the values of covariates X_{jt}. For example, if Y_{jt} is earnings of individual j in year t, the first passage time conditional on Y_{j1} and the covariates might be the earliest year in which an individual whose earnings are currently low according to some criterion becomes a high earner. X might represent characteristics of the individual such as age, education, and experience in the labor force.

Given an integer $\theta > T$, let $P(\theta \mid y_1, y^*, x)$ denote the probability that the first passage time for threshold y^* and individual i exceeds θ conditional on $Y_{j1} = y_1$ and $X_{jt} = x_t$ $(t = 1, \ldots, \theta)$. Then

$$P(\theta \mid y_1, y^*, x) = \Pr(Y_{j2} \le y^*, \ldots, Y_{j\theta} \le y^* \mid Y_{j1} = y_1).$$

To obtain a formula for $P(\theta \mid y_1, y^*, x)$, let f_W denote the probability density function of the random variable $W \equiv U + \varepsilon$, and let F_ε denote the CDF of ε. If U is independent of ε and X and if ε_{jt} is independently and identically distributed for all j and t, then some algebra shows that

$$(4.16) \quad P(\theta \mid y_1, y^*, x) =$$

$$\frac{1}{f_W(y_1 - x_{j1}\beta)} \int_{-\infty}^{\infty} f_\varepsilon(y_1 - x_1\beta) \left[\prod_{k=2}^{\theta} F_\varepsilon(y^* - x_k\beta - u) \right] f_U(u)\,du.$$

It can be seen from (4.16) that $P(\theta \mid y_1, y^*, x)$ depends on both f_U and f_ε. Thus, estimation of these densities is needed to estimate the probability distribution of first passage times.

In this discussion, the data used to estimate f_U and f_ε are assumed to consist of T observations on each of n randomly sampled individuals. Thus, the data have the form $\{Y_{jt}, X_{jt}: j = 1, \ldots, n; \ t = 1, \ldots, T\}$. Typically in panel data, the number of sampled individuals is large but the number of observations per individual is not. Thus, asymptotic results will be developed under the assumption that $n \to \infty$ while T stays fixed. It will be assumed that U is independent of ε and X. For the moment, it will also be assumed that the distribution of ε_{jt} is symmetrical about 0 for each j and t and that ε_{jt} is independently and identically distributed for all j and t. The assumption of symmetrically, independently, and identically distributed ε_{it}'s is unrealistic in many applications, however. Section 4.3 shows how this assumption can be removed.

4.2.1 Estimating f_U and f_ε

This section explains how to estimate f_U and f_ε nonparametrically. The asymptotic properties of the estimators are discussed in Section 4.2.2.

To begin, let b_n be a $n^{1/2}$-consistent estimator of β in (4.1), possibly one of the least-squares estimators described by Hsiao (1986), among others. Let W_{njt} ($j = 1, ..., n$; $t = 1, ..., T$) denote the residuals from the estimate of (4.1). That is

$$W_{njt} = Y_{jt} - X_{jt} b_n . \tag{4.17}$$

In addition, let η_{njt} ($t = 2, ..., T$) denote the residuals from the estimate of the differenced model for $Y_{jt} - Y_{j1}$. That is

$$\eta_{njt} = (Y_{jt} - Y_{j1}) - (X_{jt} - X_{j1}) b_n . \tag{4.18}$$

Observe that as $n \to \infty$, while T remains fixed, W_{njt} converges in distribution to $W = U + \varepsilon$, and η_{njt} converges in distribution to the random variable η that is distributed as the difference between two independent realizations of ε. Thus, the estimation data $\{Y_{jt}, X_{jt}\}$ provide estimates of random variables that are distributed as W and η. The data do not provide estimates of U and ε. However, the distribution of W is the convolution of the distributions of U and ε, whereas the distribution of η is the convolution of the distribution of ε with itself. Thus, estimation of f_U and of f_ε are problems in deconvolution.

The deconvolution problems involved in estimating f_U and f_ε are different from the one discussed in Section 4.1. Section 4.1 considered a problem in which the distribution of the observed random variable W is the convolution of U and another random variable whose distribution is known. In (4.1), however, the distribution of W is the convolution of the distribution of U with the distribution of a random variable ε whose distribution is unknown. The distribution of the observed random variable η is the convolution of the unknown distribution of ε with itself. Despite these differences, the approach to solving the deconvolution problems presented by (4.1) is similar to the approach used in Section 4.1, and the resulting estimators have the same slow rates of convergence.

To obtain estimators of f_U and f_ε, let ψ_W and ψ_η, respectively, denote the characteristic functions of W and η. Let ψ_U and ψ_ε denote the characteristic functions of U and ε. Then for any real τ

$$\psi_W(\tau) = \psi_U(\tau)\psi_\varepsilon(\tau)$$

and

$$\psi_\eta(\tau) = |\psi_\varepsilon(\tau)|^2.$$

Because f_ε is assumed to be symmetrical, ψ_ε is real-valued. Also assume, as in Section (4.1), that $\psi_\varepsilon(\tau) \neq 0$ for all finite τ. This implies that $\psi_\varepsilon(\tau) > 0$ for all finite τ because ψ_ε is a continuous function and $\psi_\varepsilon(0) = 1$. Therefore,

$$\psi_\varepsilon(\tau) = \psi_\eta(\tau)^{1/2},$$

and

$$\psi_U(\tau) = \frac{\psi_W(\tau)}{\psi_\eta(\tau)^{1/2}},$$

where both square roots are positive. It follows from the inversion formula for characteristic functions that for any real z,

$$f_\varepsilon(z) = \frac{1}{2\pi}\int_{-\infty}^{\infty} e^{-i\tau z}\psi_\eta(\tau)^{1/2}\,d\tau \tag{4.19}$$

and

$$f_U(u) = \frac{1}{2\pi}\int_{-\infty}^{\infty} e^{-i\tau u}\frac{\psi_W(\tau)}{\psi_\eta(\tau)^{1/2}}\,d\tau. \tag{4.20}$$

The unknown characteristic functions ψ_W and ψ_η can be estimated by

$$\psi_{nW}(\tau) = \frac{1}{nT}\sum_{j=1}^{n}\sum_{t=1}^{T}\exp(i\tau W_{njt})$$

for ψ_W and

$$\psi_{n\eta}(\tau) = \frac{1}{n(T-1)} \sum_{j=1}^{n} \sum_{t=2}^{T} \exp(i\tau\eta_{njt})$$

for ψ_η. As in Section 4.1, however, f_ε and f_U cannot be estimated by simply substituting replacing ψ_W and ψ_η with ψ_{nW} and $\psi_{n\eta}$ in (4.19) and (4.20) because the resulting integrals do not exist in general. Also as in Section 4.1, this problem can be solved by the Fourier-transform analog of kernel smoothing. To this end, let ζ be a random variable whose characteristic function ψ_ζ has support [-1,1] and satisfies other conditions that are stated below. Let $\{\nu_{n\varepsilon}\}$ and $\{\nu_{nU}\}$ be sequences of bandwidths that converge to 0 as $n \to \infty$. The smoothed estimators of f_ε and f_U are

$$(4.21) \qquad f_{n\varepsilon}(z) = \frac{1}{2\pi} \int_{-\infty}^{\infty} e^{-i\tau z} \left|\psi_{n\eta}(\tau)\right|^{1/2} \psi_\zeta(\nu_{n\varepsilon}\tau)d\tau$$

and

$$(4.22) \qquad f_{nU}(u) = \frac{1}{2\pi} \int_{-\infty}^{\infty} e^{-i\tau u} \frac{\psi_{nW}(\tau)\psi_\zeta(\nu_{nU}\tau)}{\left|\psi_{n\eta}(\tau)\right|^{1/2}} d\tau .$$

4.2.2 Large Sample Properties of $f_{n\varepsilon}$ and f_{nU}

The consistency and rates of convergence of $f_{n\varepsilon}$ and f_{nU} have been investigated by Horowitz and Markatou (1996). The results are summarized in this section. It is also shown that the centered, normalized estimators are asymptotically normally distributed.

Make the following assumptions:

Assumption P1: The distributions of ε and U are continuous and f_ε is symmetrical about 0. Moreover, f_ε and f_U are everywhere twice continuously differentiable with uniformly bounded derivatives, and ψ_ε is strictly positive everywhere.

Assumption P2: The distribution of X has bounded support.

Assumption P3: b_n satisfies $n^{1/2}(b_n - \beta) = O_p(1)$.

Assumption P4: Define $A_{n\varepsilon} = (\log n)/[n^{1/2}\psi_\varepsilon(1/\nu_{n\varepsilon})^2]$ and $B_{n\varepsilon} = 1/[n^{1/2}\nu_{n\varepsilon}\psi_\varepsilon(1/\nu_{n\varepsilon})^2]$. Define A_{nU} and B_{nU} by replacing $\nu_{n\varepsilon}$ with ν_{nU} in $A_{n\varepsilon}$ and

$B_{n\varepsilon}$. As $n \to \infty$, $\nu_{n\varepsilon} \to 0$, $\nu_{nU} \to 0$, $B_{n\varepsilon}/\nu_{n\varepsilon} \to 0$, $B_{nU}/\nu_{nU} \to 0$, $A_{n\varepsilon}/\nu_{n\varepsilon} = O(1)$, and $A_{nU}/\nu_{nU} = O(1)$.

Assumption P1 insures, among other things, that f_ε and f_U are identified. Examples of distributions with strictly positive characteristic functions are the normal, the Cauchy, and scale mixtures of these. Symmetry is not required for identification, however, as is explained in Section 4.3.2. Assumption P2 is made to avoid technical complications that arise when X has unbounded support. This assumption can always be satisfied by dropping observations with very large values of X. Assumption P3 insures that random sampling errors in the estimator of β are asymptotically negligible. All commonly used estimators of β satisfy this assumption. Finally, assumption P4 restricts the rates at which $\nu_{n\varepsilon}$ and ν_{nU} converge to 0.

The following theorem establishes uniform consistency of $f_{n\varepsilon}$ and f_{nU}.

Theorem 4.4: Let ψ_ζ be a bounded, real characteristic function with support [-1,1]. If ψ_ζ is twice differentiable in a neighborhood of 0 and assumptions P1-P4 hold, then as $n \to \infty$

$$\sup_z |f_{n\varepsilon}(z) - f_\varepsilon(z)| = O_p(\nu_{n\varepsilon}^2) + O_p(B_{n\varepsilon}/\nu_{n\varepsilon}) + o_p(A_{n\varepsilon}/\nu_{n\varepsilon})$$

and

$$\sup_u |f_{nU}(u) - f_U(u)| = O_p(\nu_{nU}^2) + O_p(B_{nU}/\nu_{nU}) + o_p(A_{nU}/\nu_{nU}) . \ \blacksquare$$

The proof of this theorem is given in Horowitz and Markatou (1996). The theorem can be explained heuristically by using arguments similar to those used to explain Theorems 4.1 and 4.2. Consider f_{nU}. Define $\psi_{n\varepsilon}(\tau) = |\psi_{n\eta}(\tau)|^{1/2}$. The only difference between (4.4) and (4.22) is that the denominator of the integrand of (4.22) is an estimator of ψ_ε instead of the true ψ_ε. Therefore, to obtain the conclusion of Theorem (4.4) for f_{nU}, it suffices to show that if assumptions P1-P4 hold, then an asymptotically negligible error is made by replacing $\psi_{n\varepsilon}(\tau)$ with $\psi_\varepsilon(\tau)$ in (4.22). By using an extension of the delta method, it can be shown that

$$(4.23) \quad |\psi_{n\varepsilon}(\tau) - \psi_\varepsilon(\tau)| = \psi_\varepsilon(\tau)\left[\frac{\psi_{n\eta}(\tau) - \psi_\eta(\tau)}{\psi_\eta(\tau)}\right]\left[\frac{1}{2} + o_p(1)\right]$$

and that

$$\sup_{|\tau|\le 1/\nu_{nU}} |\psi_{n\eta}(\tau)-\psi_{\eta}(\tau)| = o_p\left(\frac{\log n}{n^{1/2}}\right)+O_p\left(\frac{1}{n^{1/2}\nu_{nU}}\right). \tag{4.24}$$

See Horowitz and Markatou (1996) for details. Substituting (4.24) into (4.23) yields

$$\sup_{|\tau|\le 1/\nu_{nU}} |\psi_{n\varepsilon}(\tau)-\psi_{\varepsilon}(\tau)| = o_p(A_{nU})+O_P(B_{nU})$$

and

$$\sup_{|\tau|\le 1/\nu_{nU}} \left|\frac{\psi_{n\varepsilon}(\tau)}{\psi_{\varepsilon}(\tau)}-1\right| = o_p(A_{nU})+O_P(B_{nU}). \tag{4.25}$$

A further application of the delta method to the right-hand side of (4.22) shows that

$$\frac{1}{2\pi}\int_{-\infty}^{\infty} e^{-i\tau u}\frac{\psi_{nW}(\tau)\psi_{\zeta}(\nu_{nU}\tau)}{\psi_{n\varepsilon}(\tau)}d\tau \tag{4.26}$$

$$=\frac{1}{2\pi}\int_{-\infty}^{\infty} e^{-i\tau u}\frac{\psi_{nW}(\tau)\psi_{\zeta}(\nu_{nU}\tau)}{\psi_{\varepsilon}(\tau)}d\tau + R_{nU},$$

where $R_{nU}=O_p(A_{nU})$ is an asymptotically negligible remainder term. Therefore, it suffices to investigate the convergence of the first term on the right-hand side of (4.26). This term, however, is the same as the right-hand side of (4.4), so the arguments used with (4.4) also apply to (4.22).

The rate of convergence of $f_{n\varepsilon}$ can be obtained by writing (4.21) in the form

$$f_{n\varepsilon}(z)=\frac{1}{2\pi}\int_{-\infty}^{\infty} e^{-i\tau z}\frac{\psi_{n\eta}(\tau)\psi_{\zeta}(\nu_{n\varepsilon}\tau)}{\psi_{n\varepsilon}(\tau)}d\tau. \tag{4.27}$$

The arguments leading to (4.25) show that (4.25) continues to hold if ν_{nU} is replaced by $\nu_{n\varepsilon}$, and A_{nU} and B_{nU} are replaced by $A_{n\varepsilon}$ and $B_{n\varepsilon}$. By applying this modified form of (4.25) to the right-hand side of (4.27), it can be shown that

$$\frac{1}{2\pi}\int_{-\infty}^{\infty} e^{-i\tau z}\frac{\psi_{n\eta}(\tau)\psi_{\zeta}(\nu_{n\varepsilon}\tau)}{\psi_{n\varepsilon}(\tau)}d\tau \tag{4.28}$$

$$= \frac{1}{2\pi} \int_{-\infty}^{\infty} e^{-i\tau z} \frac{\psi_{n\eta}(\tau)\psi_{\varsigma}(\nu_{n\varepsilon}\tau)}{\psi_{\varepsilon}(\tau)} d\tau + R_{n\varepsilon},$$

where $R_{n\varepsilon} = O_p(A_{n\varepsilon})$ is an asymptotically negligible remainder term. Therefore, the asymptotic properties of $f_{n\varepsilon}$ are the same as those of the first term on the right-hand side of (4.28). This term is the same as the right-hand side of (4.4) except with $\psi_{n\eta}$ in place of ψ_{nW}. Therefore, the arguments used with (4.4) can also be used to analyze $f_{n\varepsilon}$.

As in the deconvolution problem of Section 4.1, the rates of convergence of $f_{n\varepsilon}$ and f_{nU} are controlled by the rates at which the bandwidths $\nu_{n\varepsilon}$ and ν_{nU} converge to 0. These, in turn, are controlled by the thickness of the tails of ψ_{ε}, as can be seen from assumption P4. Faster rates of convergence are possible when the tails are thick than when they are thin. Horowitz and Markatou (1996) investigate in detail the case of $\varepsilon \sim N(0,\sigma^2)$. They show that when f_{ε} and f_U are assumed to be twice differentiable, the fastest possible rate of convergence of an estimator of either density is $(\log n)^{-1}$. Under the assumptions of Theorem 4.4, this rate is achieved by $f_{n\varepsilon}$ and f_{nU} in (4.20) and (4.21).

Equations (4.26) and (4.28) can also be used to show that the standardized versions of $f_{n\varepsilon}$ and f_{nU} are asymptotically normally distributed. Let $g_{nU}(u)$ and $g_{n\varepsilon}(z)$, respectively, denote the first terms on the right-hand sides of (4.26) and (4.28). When f_{nU} and $f_{n\varepsilon}$ have their fastest rates of convergence, their rates of convergence and variances are $O(\nu_{nU}^2)$ and $O(\nu_{n\varepsilon}^2)$. Denote these rates by $\rho_U(n)$ and $\rho_{\varepsilon}(n)$. Then

$$\rho_{\varepsilon}(n)^{-1}[f_{n\varepsilon}(z) - f_{\varepsilon}(z)] = \rho_{\varepsilon}(n)^{-1}[g_{n\varepsilon}(z) - g_{\varepsilon}(z)] + \rho_{\varepsilon}(n)^{-1} R_{n\varepsilon}$$

and

$$\rho_U(n)^{-1}[f_{nU}(u) - f_{\varepsilon U}(u)] = \rho_U(n)^{-1}[g_{nU}(u) - g_U(u)] + \rho_U(n)^{-1} R_{nU}.$$

Therefore, $\rho_{\varepsilon}(n)^{-1}[f_{n\varepsilon}(z) - f_{\varepsilon}(z)]$ is asymptotically equivalent to $\rho_{\varepsilon}(n)^{-1}[g_{n\varepsilon}(z) - f_{\varepsilon}(z)]$ and the asymptotic normality results of Section 4.1.3 can be applied to $f_{n\varepsilon}(z)$ if $\rho_{\varepsilon}(n)^{-1}R_{n\varepsilon} = o_p(1)$ as $n \to \infty$. Similarly, the asymptotic normality results of Section 4.1.2 apply to $f_{nU}(u)$ if $\rho_U(n)^{-1}R_{nU} = o_p(1)$. When $f_{n\varepsilon}$ and f_{nU} have their fastest possible rates of convergence, $\rho_{n\varepsilon}(n) = O(\nu_{n\varepsilon}^2)$ and $\rho_{nU}(n) = O(\nu_{nU}^2)$. In addition $R_{n\varepsilon} = O_p(A_{n\varepsilon})$, and $R_{nU} = O_p(A_{nU})$. Therefore, if $\nu_{n\varepsilon}$ and ν_{nU} are chosen to optimize the rates of convergence of $f_{n\varepsilon}$ and f_{nU}, the asymptotic normality results of Section 4.1.2 apply to $f_{n\varepsilon}(z)$ and $f_{nU}(u)$ provided that assumption P4 is strengthened to require $A_{n\varepsilon}/\nu_{n\varepsilon}^2 \to 0$ and $A_{nU}/\nu_{nU}^2 \to 0$ as $n \to \infty$

4.2.3 Estimating First Passage Times

This section shows how to estimate $P(\theta | y_1, y^*, x)$, the probability distribution of first passage times conditional on the covariates and the initial value of Y when Y_{it} follows model (4.1). An application of the estimator is presented in Section 4.4.

Equation (4.16) provides a formula for $P(\theta | y_1, y^*, x)$. The right-hand side of (4.16) can be estimated consistently by replacing β with b_n, f_ε with $f_{n\varepsilon}$, f_U with f_{nU}, f_W with a kernel estimator of the density of the residuals W_{njt}, and F_ε with a consistent estimator. One way to obtain an estimator of F_ε is by integrating $f_{n\varepsilon}$. The resulting estimator is

$$F_{n\varepsilon}(z) = \int_{-M_n}^{z} f_{n\varepsilon}(\xi)d\xi ,$$

where $M_n \to \infty$ at a suitable rate as $n \to \infty$. This estimator has the practical disadvantage of requiring selection of the tuning parameter M_n.

The need to choose the additional tuning parameter M_n can be avoided by taking advantage of the symmetry of the distribution of ε. Symmetry implies that $F_\varepsilon(z) = 1 - F_\varepsilon(-z)$. Therefore, $F_\varepsilon(z) = 0.5 + 0.5[F_\varepsilon(z) - F_\varepsilon(-z)]$. But

$$(4.29) \qquad 0.5 + 0.5[F_\varepsilon(z) - F_\varepsilon(-z)] = 0.5 + 0.5\int_{-z}^{z} f_\varepsilon(z)dz .$$

In addition,

$$(4.30) \qquad f_\varepsilon(z) = \frac{1}{2\pi}\int_{-\infty}^{\infty} e^{-iz\tau}\psi_\varepsilon(\tau)d\tau .$$

Therefore, substituting (4.30) into (4.29) yields

$$(4.31) \qquad F_\varepsilon(z) = 0.5 + \frac{1}{\pi}\int_{-\infty}^{\infty} \frac{\sin z\tau}{\tau}\psi_\varepsilon(\tau)d\tau .$$

But $(\sin z\tau)/\tau$ is an even function, and symmetry of f_ε implies that ψ_ε also is even. Therefore, (4.31) can be written in the form

$$(4.32) \qquad F_\varepsilon(z) = 0.5 + \frac{2}{\pi}\int_{0}^{\infty} \frac{\sin z\tau}{\tau}\psi_\varepsilon(\tau)d\tau$$

Equation (4.32) forms the basis of the estimator of F_ε that is proposed here. Using arguments similar to those applicable to Theorem 4.4, it may be shown that $F_\varepsilon(z)$ is estimated consistently by

$$(4.33) \qquad F_{n\varepsilon}(z) = 0.5 + \frac{2}{\pi}\int_0^\infty \frac{\sin z\tau}{\tau}\left|\psi_\eta(\tau)\right|^{1/2}\psi_\varsigma(\nu_{n\varepsilon}\tau)d\tau$$

The estimator of the right-hand side of (4.16) is completed by replacing F_ε with $F_{n\varepsilon}$ from (4.33). The result is that $P(\theta\,|y_1, y^*, x)$ is estimated consistently by

$$(4.34) \qquad P_n(\theta \mid y_1, y^*, x) =$$

$$\frac{1}{f_{nW}(y_1 - x_{j1}b_n)}\int_{-m_n}^{m_n} f_{n\varepsilon}(y_1 - x_1 b_n)\left[\prod_{k=2}^{\theta} F_{n\varepsilon}(y^* - x_k b_n - u)\right] f_{nU}(u)du\,,$$

where f_{nW} is a kernel estimator of the density of W_{njt} and $\{m_n\}$ is a sequence of positive constants that satisfies $m_n \to \infty$,

$$\sup_z m_n\left|f_{n\varepsilon}(z) - f_\varepsilon(z)\right| \to^p 0$$

and

$$\sup_u m_n\left|f_{nU}(u) - f_U(u)\right| \to^p 0$$

as $n \to \infty$.

4.2.4 Bias Reduction

As was discussed in Section 4.2.1, $f_{n\varepsilon}$ and f_{nU} are asymptotically equivalent to the density estimator (4.4). Therefore, it follows from (4.11) that $f_{n\varepsilon}$ and f_{nU} have biases of sizes $O(\nu_{n\varepsilon}^2)$ and $O(\nu_{nU}^2)$, respectively. The results of Monte Carlo experiments reported by Horowitz and Markatou (1996) and summarized in Section 4.2.4 show that this bias can have a large effect on the accuracy of $P_n(\theta\,|y_1, y^*, x)$ as an estimator of $P(\theta\,|y_1, y^*, x)$. This section describes corrections for $f_{n\varepsilon}$ and f_{nU} that remove parts of these biases. The Monte Carlo experiments reported in Section 4.2.4 show that the accuracy of $P_n(\theta\,|y_1, y^*, x)$ is greatly increased when the corrected density estimators are used. The arguments leading to the corrections for $f_{n\varepsilon}$ and f_{nU} are identical, so only $f_{n\varepsilon}$ is discussed here.

To derive the correction for $f_{n\varepsilon}$, write $f_{n\varepsilon}(z) - f_\varepsilon(z)$ in the form

$$f_{n\varepsilon}(z) - f_\varepsilon(z) = \Delta_{n1}(z) + \Delta_{n2}(z),$$

where

$$\Delta_{n1}(z) = \frac{1}{2\pi}\int_{-\infty}^{\infty} e^{-i\tau z}\left[\left|\psi_{n\eta}(\tau)\right|^{1/2} - \psi_\varepsilon(\tau)\right]\psi_\zeta(\nu_{n\varepsilon}\tau)d\tau$$

and

$$\Delta_{n2}(z) = \frac{1}{2\pi}\int_{-\infty}^{\infty} e^{-i\tau z}[\psi_\zeta(\nu_{n\varepsilon}\tau) - 1]\psi_\varepsilon(\tau)d\tau .$$

The quantity $\Delta_{n2}(z)$ is nonstochastic. In a finite sample, neither $E\Delta_{n1}(z)$ nor $\Delta_{n2}(z)$ is zero in general, so $f_{n\varepsilon}(z)$ is biased. Of course, the biases vanish as $n \to \infty$, as is needed for consistency of $f_{n\varepsilon}(z)$. $E\Delta_{n1}(z)$ is the component of bias caused by estimating ψ_ε, and $\Delta_{n2}(z)$ is the component of bias caused by smoothing the empirical distribution of η. The correction described in this section removes the second component of bias through $O(\nu_{n\varepsilon}^2)$.

The correction is obtained by proceeding in a manner similar to that used to derive (4.11). The quantity $\psi_\zeta(\nu_{n\varepsilon}\tau)\psi_\varepsilon(\tau)$ is the characteristic function of the random variable $\varepsilon + \nu_{n\varepsilon}\zeta$. Therefore, $\Delta_{n2}(z)$ is the difference between the probability densities of $\varepsilon + \nu_{n\varepsilon}\zeta$ and ε. Let f_ζ denote the probability density function of ζ. Then

$$\Delta_{n2}(z) = \int_{-\nabla}^{\infty} f_\varepsilon(z - \nu_{n\varepsilon}\tau) f_\zeta(\tau)d\tau - f_\varepsilon(z).$$

A Taylor series expansion of $f_\varepsilon(z - \nu_{n\varepsilon}\tau)$ about $\nu_{n\varepsilon} = 0$ yields

$$(4.35) \qquad \Delta_{n2}(z) = \frac{1}{2}\nu_{n\varepsilon}^2 f_\varepsilon''(z)\sigma_\zeta^2 + o(\nu_{n\varepsilon}^2),$$

where σ_ζ^2 is the variance of ζ. The first term on the right-hand side of (4.35) is the smoothing bias in $f_{n\varepsilon}(z)$ through $O(\nu_{n\varepsilon}^2)$.

Note that σ_ζ^2 is known because ψ_ζ and, therefore, f_ζ is chosen by the analyst. Let $f_{n\varepsilon}''(z)$ be a consistent estimator of $f_\varepsilon''(z)$. Then the $O(\nu_{n\varepsilon}^2)$ smoothing bias in $f_{n\varepsilon}(z)$ can be removed by estimating f_ε with

$$\hat{f}_{n\varepsilon}(z) = f_{n\varepsilon}(z) - \nu_{n\varepsilon}^2 f_{n\varepsilon}''(z)\sigma_\varsigma^2 .$$

A consistent estimator of $f_\varepsilon''(z)$ can be obtained by differentiating $f_{n\varepsilon}(z)$. As is normal in kernel estimation, however, estimating the derivative requires using a bandwidth that converges more slowly than the bandwidth that is used for estimating $f_\varepsilon(z)$ itself. The formal result is stated in the following theorem, which is proved in Horowitz and Markatou (1996).

Theorem 4.5: Let assumptions P1-P4 hold. Assume that f_ε'' is Lipschitz continuous. That is $|f_\varepsilon''(z + \delta) - f_\varepsilon''(z)| \leq c\delta$ for any z, all sufficiently small $\delta > 0$, and some $c > 0$. Let $\{\gamma_{n\varepsilon}\}$ be a positive sequence satisfying $\gamma_{n\varepsilon} \to 0$, $B_{n\varepsilon}/\gamma_{n\varepsilon}^2 \to 0$, and $A_{n\varepsilon}/\gamma_{n\varepsilon} = O(1)$ as $n \to \infty$. Define

$$f_{n\varepsilon}''(z) = -\frac{1}{2\pi}\int_{-\infty}^{\infty} e^{-iz\tau}\tau^2 \left|\psi_{n\eta}(\tau)\right|^{1/2} \psi_\varsigma(\gamma_{n\varepsilon}\tau)d\tau$$

Then

$$\operatorname*{plim}_{n\to\infty} \sup_z \left| f_{n\varepsilon}''(z) - f_\varepsilon''(z) \right| = 0 . \blacksquare$$

The procedure for removing smoothing bias from $f_{nU}(u)$ is similar. The resulting estimator is

$$\hat{f}_{nU}(u) = f_{nU}(u) - \nu_{nU}^2 f_{nU}''(u)\sigma_\varsigma^2 ,$$

where $f_{nU}''(u)$ is the following consistent estimator of $f_U''(u)$:

$$f_{nU}''(u) = -\frac{1}{2\pi}\int_{-\infty}^{\infty} e^{-iu\tau}\tau^2 \psi_{nW}(\tau)\psi_\varsigma(\gamma_{nU}\tau)d\tau ,$$

where $\{\gamma_{nU}\}$ is a positive sequence that converges to zero sufficiently slowly. The following modified version of Theorem 4.5 applies to $f_{nU}''(u)$:

Theorem 4.5′: Let assumptions P1-P4 hold. Assume that f_U'' is Lipschitz continuous. That is $|f_U''(u + \delta) - f_U''(u)| \leq c\delta$ for any u, all sufficiently small $\delta > 0$, and some $c > 0$. Let $\{\gamma_{nU}\}$ be a positive sequence satisfying $\gamma_{nU} \to 0$, $B_{nU}/\gamma_{nU}^2 \to 0$, and $A_{nU}/\gamma_{nU} = O(1)$ as $n \to \infty$. Then

$$\operatorname*{plim}_{n\to\infty} \sup_u \left| f_{nU}''(u) - f_U''(u) \right| = 0 . \blacksquare$$

4.2.5 Monte Carlo Experiments

Horowitz and Markatou (1996) carried out a Monte Carlo investigation of the ability of $P_n(\theta | y_1, y^*, x)$ to provide useful information about $P(\theta | y_1, y^*, x)$ with samples of moderate size. Data for the experiments were generated by simulation from the model

$$Y_{jt} = U_j + \varepsilon_{jt}, \quad j = 1, \ldots, 1000; t = 1,2.$$

Thus, the simulated data correspond to a panel of length $T = 2$ composed of $n = 1000$ individuals. There were no covariates in the experiments. The distribution of U was N(0,1). The distribution of ε was N(0,1) in one experiment. In the other, ε was sampled from N(0,1) with probability 0.9 and from N(0,16) with probability 0.1. This mixture distribution has tails that are thicker than the tails of the normal distribution. In both sets of experiments, the fastest possible rate of convergence in probability of estimators of f_ε and f_U is $(\log n)^{-1}$, so the experiments address situations in which the estimators converge slowly.

As in the experiments reported in Section 4.1.3, the smoothing function ψ_ζ was the fourfold convolution of the uniform density with itself. The density f_ν was estimated using a kernel estimator with the standard normal density function as the kernel. In estimating $P(\theta | y_1, y^*)$, $y_1 = -1$, $y^* = 1$, and $\theta = 3, 5, 7, 9, 11$. The bandwidths were set at values such that $\psi_{n\varepsilon}(\lambda_{n\varepsilon}^{-1})$ and $\psi_{nU}(\lambda_{nU}^{-1})$ were both approximately zero. There were 100 Monte Carlo replications per experiment.

The results are shown in Table 4.1. Columns 3-5 present the true values of $P(\theta | y_1 = -1, y^* = 1)$, the means of the Monte Carlo estimates of $P(\theta | y_1 = -1, y^* = 1)$ that were obtained using $f_{n\varepsilon}$ and f_{nU} without bias correction, and the means of the Monte Carlo estimates that were obtained using the bias-corrected forms of $f_{n\varepsilon}$ and f_{nU}. The estimates of $P(\theta | y_1 = -1, y^* = 1)$ that were obtained without

Table 4.1 Results of Monte Carlo Experiments with Estimator $P(\theta | y_1, y^*)$

Distr. of ε	θ	True Prob.	With Bias Corr.	Without Bias Corr.	Assuming Normal ε
Normal	3	0.89	0.88	0.78	
	5	0.81	0.77	0.69	
	7	0.74	0.69	0.62	
	9	0.69	0.62	0.56	
	11	0.64	0.57	0.51	
Mixture	3	0.86	0.83	0.74	0.76
	5	0.76	0.71	0.73	0.60
	7	0.67	0.61	0.55	0.49
	9	0.60	0.54	0.49	0.41
	11	0.55	0.48	0.44	0.35

Source: Horowitz and Markatou (1996).

using bias correction for $f_{n\varepsilon}$ and f_{nU} are biased downward by 12-20 percent, depending on the distribution of ε and the value of θ. Using the bias-corrected density estimates, however, reduces the downward bias of $P_n(\theta \mid y_1 = -1, y^* = 1)$ to 1-13 percent. Thus, the bias correction removes 35 percent to virtually all of the bias of $P_n(\theta \mid y_1 = -1, y^* = 1)$, depending on the distribution of ε and the value of θ. This illustrates the usefulness of the bias-correction procedure.

The last column of Table 4.1 shows the means of the estimates of $P(\theta \mid y_1 = -1, y^* = 1)$ that were obtained by assuming that ε is normally distributed when, in fact, it has the mixture distribution N(0,1) with probability 0.9 and N(0,16) with probability 0.1. This is an important comparison because normality is often assumed in applications. It can be seen that the erroneous assumption of normality produces estimates that are biased downward by 12-36 percent, whereas the downward bias is only 1-13 percent when the bias-corrected nonparametric density estimators are used. Although not shown in the table, the Monte Carlo results also reveal that use of the nonparametric density estimators reduces the mean-square error of $P_n(\theta \mid y_1 = -1, y^* = 1)$ as well as the bias.

There is a simple intuitive explanation for the severe downward bias of the parametric estimator of $P(\theta \mid y_1 = -1, y^* = 1)$. The mixture distribution used in the experiment has less probability in its tails than does a normal distribution with the same variance. Therefore, the normal distribution has a higher probability of a transition from one tail to another than does the mixture distribution. Since $P(\theta \mid y_1 = -1, y^* = 1)$ is the probability that a transition between the tails does not occur, the probabilities obtained from the normal distribution are too low. In summary, the Monte Carlo evidence indicates that the nonparametric estimation procedure with bias correction can yield estimates of first-passage probabilities that are considerably more accurate than the ones obtained from a misspecified parametric model.

4.3 Extensions

The discussion to this point has assumed that the ε_{jt} are independently and symmetrically distributed. These assumptions are unrealistic in many applications. They are relaxed in this section. Sections 4.3.1 to 4.3.3 treat the case of serially correlated ε_{jt} that are symmetrically distributed. Asymmetrically but independently distributed ε_{jt}'s are treated in Section 4.3.4. The case of both serial correlation and asymmetry is treated in Section 4.3.5.

4.3.1 Serially Correlated ε's

In this section and in Sections 4.3.2 and 4.3.3, it is assumed that for each j, $\{\varepsilon_{jt}: t = 1, 2,...\}$ follows either an autoregressive or moving average process. The process is assumed to be the same for each j, and the ε_{jt} are assumed to be identically and symmetrically distributed for each j and t. The main problem to be solved is estimating ψ_ε and f_ε. Given an estimator $\psi_{n\varepsilon}$ of ψ_ε, f_U can be estimated by using the following slightly modified version of (4.22):

$$(4.36) \qquad f_{nU}(u) = \frac{1}{2\pi} \int_{-\infty}^{\infty} e^{-i\tau u} \frac{\psi_{nW}(\tau)\psi_\zeta(\nu_{nU}\tau)}{\psi_{n\varepsilon}(\tau)} d\tau .$$

To begin, suppose that ε_{jt} follows a stationary AR(1) process and that $T \geq 3$. Higher-order AR processes are treated in Section 4.3.2. MA processes are treated in Section 4.3.3.

If ε_{jt} follows an AR(1) process, then

$$(4.37) \qquad \varepsilon_{jt} = \alpha\varepsilon_{j,t-1} + \xi_{jt}$$

for each j, where $|\alpha| < 1$ and ξ_{jt} is independently and identically distributed across j and t. Define $\eta_{jt} = \varepsilon_{jt} - \varepsilon_{j,t-1}$. Then (4.37) implies that

$$\eta_{jt} = -(1-\alpha)\varepsilon_{j,t-1} + \xi_{jt}$$

But $\varepsilon_{j,t-1}$ is independent of ξ_{jt}. Therefore, the characteristic functions of η, ε, and ξ are related by

$$\psi_\eta(\tau) = \psi_\varepsilon[-(1-\alpha)\tau]\psi_\xi(\tau)$$

$$(4.38) \qquad = \psi_\varepsilon[(1-\alpha)\tau]\psi_\xi(\tau)$$

where (4.38) follows from the symmetry of the distribution of ε_{jt}. Solving (4.38) for ψ_ε yields

$$\psi_\varepsilon(\tau) = \frac{\psi_\eta[\tau/(1-\alpha)]}{\psi_\xi[\tau/(1-\alpha)]} .$$

To estimate ψ_ε, let α_n be a consistent estimator of α. Methods for estimating α are described by Hsiao (1986). Let η_{njt} be the residual from least-squares estimation of

$$Y_{jt} - Y_{j,t-1} = (X_{jt} - X_{j,t-1})\beta + \varepsilon_{jt} - \varepsilon_{j,t-1} .$$

Let $\psi_{n\eta}$ be the empirical characteristic function of η_{njt}:

(4.39) $$\psi_{n\eta}(\tau) = \frac{1}{n(T-1)} \sum_{j=1}^{n} \sum_{t=2}^{T} \exp(i\tau\eta_{njt}).$$

If $\psi_{n\xi}$ is a consistent estimator of ψ_ξ, the characteristic function of ξ,, then $\psi_\varepsilon(\tau)$ is estimated consistently by

$$\psi_{n\varepsilon}(\tau) = \frac{\psi_{n\eta}[\tau/(1-\alpha_n)]}{\psi_{n\xi}[\tau/(1-\alpha_n)]}.$$

It follows that f_ε and the density of ξ, f_ξ, can be estimated by substituting $\psi_{n\varepsilon}$ and $\psi_{n\xi}$, respectively, into (4.21) in place of $|\psi_{n\eta}|^{1/2}$.

To estimate ψ_ξ, observe that

(4.40) $$\xi_{jt} - \xi_{j,t-1} = (Y_{jt} - Y_{j,t-1}) - \alpha(Y_{j,t-1} - Y_{j,t-2})$$
$$- [(X_{jt} - X_{j,t-1}) - \alpha(X_{j,t-1} - X_{j,t-2})]\beta.$$

The characteristic function of the right-hand side of (4.40) is $|\psi_\xi|^2$. Let $|\psi_{n\xi}|^2$ be the empirical characteristic function of the random variable δ_{njt} that is obtained from (4.40) by replacing α and β with consistent estimators. That is,

$$|\psi_{n\xi}(\tau)|^2 = \frac{1}{n(T-2)} \sum_{j=1}^{n} \sum_{t=3}^{T} \exp(i\tau\delta_{njt}).$$

Then ψ_ξ is estimated consistently by $(|\psi_{n\xi}|^2)^{1/2}$.

4.3.2 AR(p) Processes

Now let ε_{jt} follow a stationary AR(p) process ($p < \infty$) for each j, and assume that $T \geq p + 2$. Then

$$\varepsilon_{jt} = \sum_{k=1}^{p} a_k \varepsilon_{j,t-k} + \xi_{jt}$$

for each j, where ξ_{jt} is independently and identically distributed across j and t, and

$$\sum_{k=1}^{p} |\alpha_k| < 1.$$

To obtain an equation for ψ_ε in terms of estimable quantities, define $\eta_{jk} = \varepsilon_{j,t} - \varepsilon_{j,t-k}$ ($k = 1, ..., p$). Let η_j^* be the vector $(\eta_{j1}, \eta_{j2}, ..., \eta_{jp})'$, ε_j^* be the vector $(\varepsilon_{j,t-1}, ..., \varepsilon_{j,t-p})'$, and g be a $p\times 1$ vector of ones. Then it is easily shown that

$$\eta_j^* = A\varepsilon_j^* + g\xi_{jt}, \tag{4.41}$$

where

$$A = \begin{bmatrix} \alpha_1 - 1 & \alpha_2 & \ldots & \alpha_p \\ \alpha_1 & \alpha_2 - 1 & \ldots & \alpha_p \\ & \ldots & \ldots & \\ \alpha_1 & \alpha_2 & \ldots & \alpha_p - 1 \end{bmatrix}. \tag{4.42}$$

Solving (4.41) for ε_j^* yields

$$\varepsilon_j^* = A^{-1}\eta_j^* - A^{-1}g\xi_{jt}. \tag{4.43}$$

Equations (4.43) are a set of p equations for $\varepsilon_{j,t-k}$ ($k = 1, ..., p$). It is possible, at least in principle, to use all of these equations to estimate ψ_ε. Here, however, only the equation for $\varepsilon_{j,t-1}$ will be used. Let a denote the first row of A^{-1}. Note that a is a $1\times p$ vector. Let $c = A^{-1}g$, and note that c is a scalar. Then the first row of (4.43) can be written

$$\varepsilon_{j,t-1} = a\eta_j^* - c\xi_{jt}. \tag{4.44}$$

Let ψ_{η^*} denote the characteristic function of η^*. Then, because ξ_{jt} is symmetrically distributed and independent of the components of η_j^*, it follows from (4.44) that

$$\psi_\varepsilon(\tau) = \frac{\psi_{\eta^*}(\tau)}{\psi_\xi(c\tau)}. \tag{4.45}$$

$\psi_\varepsilon(\tau)$ can now be estimated by replacing $\psi_{\eta^*}(\tau)$ and $\psi_\xi(c\tau)$ with consistent estimators on the right-hand side of (4.45).

To estimate ψ_{η^*}, let a_n be a $n^{1/2}$-consistent estimator of a (e.g., one that is obtained by substituting $n^{1/2}$-consistent estimators of $\alpha_1, ..., \alpha_p$ into (4.42)). Let η_{njtk} be the residual from least-squares estimation of

$$Y_{jt} - Y_{j,t-k} = (X_{jt} - X_{j,t-k})\beta + \varepsilon_{jt} - \varepsilon_{j,t-k}. \tag{4.46}$$

Let η_{njt}^* be the vector $(\eta_{njt1}, \ldots, \eta_{njtp})'$. Then $\psi_{\eta^*}(\tau)$ is estimated consistently by the empirical characteristic function of $a_n\eta_{njt}^*$. The estimator is

$$\psi_{n\tau\eta^*}(\tau) = \frac{1}{n(T-p-1)} \sum_{j=1}^{n} \sum_{t=p+1}^{T} \exp(i\tau a_n \eta_{njt}).$$

To estimate ψ_ξ, observe that

$$\xi_{jt} - \xi_{j,t-1} = (Y_{jt} - Y_{j,t-1}) - \sum_{k=1}^{p} (Y_{j,t-k} - Y_{j,t-k-1}) \tag{4.47}$$
$$- \left[(X_{jt} - X_{j,t-1}) - \sum_{k-1}^{p} \alpha_k (X_{j,t-k} - X_{j,t-k-1}) \right] \beta.$$

The characteristic function of the right-hand side of (4.47) is $|\psi_\xi|^2$. Let $|\psi_{n\xi}|^2$ be the empirical characteristic function of the random variable δ_{njt} that is obtained from (4.47) by replacing the α_k and β with consistent estimators. That is,

$$|\psi_{n\xi}(\tau)|^2 = \frac{1}{n(T-p-1)} \sum_{j=1}^{n} \sum_{t=p+2}^{T} \exp(i\tau\delta_{njt}).$$

Then ψ_ξ is estimated consistently by $(|\psi_{n\xi}|^2)^{1/2}$.

4.3.3 MA Processes

It is also possible to estimate f_ε under the assumption that ε_{jt} follows a moving average process. Suppose, to begin, that ε_{jt} $(t = 1, \ldots, T)$ follows an MA(1) process. Then $\varepsilon_{jt} = \xi_{jt} + \alpha\xi_{j,t-1}$ for each j, where the random variables ξ_{jt} are independently and identically distributed across j and t. Also, assume that $T \geq 3$ and that the density of the ξ_{jt} is symmetrical about 0. As in the autoregressive case, f_U can be estimated using (4.22), so it is necessary only to find estimators of f_ε and f_ξ. To do this, suppose for the moment that $\alpha \neq 1$. Then

$$\psi_\varepsilon(\tau) = \psi_\xi(\tau)\psi_\xi(\alpha\tau).$$

Let α_n be a consistent estimator of α, possibly a method-of-moments estimator based on the residuals from least-squares estimation of (4.46). Let $\psi_{n\xi}(\tau)$ be a consistent estimator of $\psi_\xi(\tau)$. Then $\psi_\varepsilon(\tau)$ is estimated consistently by

$$\psi_{n\varepsilon}(\tau) = \psi_{n\xi}(\tau)\psi_{n\xi}(\alpha_n \tau) .$$

It follows that f_ε and f_ξ, can be estimated by substituting $\psi_{n\varepsilon}$ and $\psi_{n\xi}$, respectively, into (4.21) in place of $|\psi_{n\eta}|^{1/2}$.

To obtain an estimator of $\psi_\xi(\tau)$, define $\eta_{jt1} = \varepsilon_{jt} - \varepsilon_{j,t-1}$ and $\eta_{jt2} = \varepsilon_{jt} - \varepsilon_{j,t-2}$. Then

$$\eta_{jt1} = \xi_{jt} - (1-\alpha)\xi_{j,t-1} - \alpha\xi_{j,t-2}$$

and

$$\eta_{jt2} = \xi_{jt} + \alpha\xi_{j,t-1} - \xi_{j,t-2} - \alpha\xi_{j,t-3} .$$

Let $\psi_{\eta k}$ denote the characteristic function of η_{jtk} ($k = 1,2$). Then, because ξ_{jt} and $\xi_{j,t-k}$ are independent if $k \neq 0$,

$$\psi_{\eta 1}(\tau) = \psi_\xi(\tau)\psi_\xi[(1-\alpha)\tau]\psi_\xi(\alpha\tau)$$

and

$$\psi_{\eta 2}(\tau) = \left|\psi_\xi(\tau)\right|^2 \left|\psi_\xi(\alpha\tau)\right|^2 .$$

Therefore,

$$\psi_\varepsilon(\tau) = \left|\psi_{\eta 2}(\tau)\right|^{1/2} ,$$

and

$$\text{(4.49)} \qquad \psi_\xi(\tau) = \frac{\psi_{n1}[\tau/(1-\alpha)]}{\left|\psi_{n2}[\tau/(1-\alpha)]\right|^{1/2}} .$$

$\psi_\varepsilon(\tau)$ and $\psi_\xi(\tau)$ can be estimated by replacing $\psi_{\eta 1}(\tau)$ and $\psi_{\eta 2}(\tau)$ with consistent estimators in (4.48) and (4.49). $\psi_{\eta 1}(\tau)$ is estimated consistently by the right-hand side of (4.39). The estimator of $\psi_{\eta 2}$ can be set equal to the empirical characteristic function of the residuals from least-squares estimation of

$$Y_{jt} - Y_{j,t-2} = (X_{jt} - X_{j,t-2})\beta + \varepsilon_{jt} - \varepsilon_{j,t-2}$$

This yields the estimator

$$\psi_{n\eta 2}(\tau) = \frac{1}{n(T-2)} \sum_{j=1}^{n} \sum_{t=3}^{T} \exp(i\tau\eta_{njt2}),$$

where η_{njt2} denotes the residual.

If $\alpha = 1$, then $\psi_{\eta 1}(\tau) = \psi_\varepsilon(\tau)$. Therefore, ψ_ε can be estimated from (4.39) if $\alpha = 1$.

Now suppose that ε_{jt} follows an MA(q) process ($q < \infty$) for each j. Then

$$\varepsilon_{jt} = \xi_{jt} + \sum_{k=1}^{q} \alpha_k \xi_{j,t-k}$$

for each j, where α_k ($k = 1, \ldots, q$) is a constant and the ξ_{jt} are independently and identically distributed across j and t. Let $T \geq q + 2$, and assume that the MA process for ε_{jt} is invertible. Define $\eta_{jt,q+1} = \varepsilon_{jt} - \varepsilon_{j,t-q-1}$. Then

$$\eta_{jt,q+1} = \xi_{jt} + \sum_{k=1}^{q} \alpha_k \xi_{j,t-k} - \xi_{j,t-q-1} - \sum_{k=1}^{q} \alpha_k \xi_{j,t-q-k-1}.$$

The characteristic function of $\eta_{jt,q+1}$ satisfies

$$\psi_{\eta,q+1}(\tau) = |\psi_\xi(\tau)|^2 \prod_{j=1}^{q} |\psi_\xi(\alpha_j \tau)|^2$$

$$= |\psi_\varepsilon(\tau)|^2.$$

Therefore,

$$\psi_\varepsilon(\tau) = |\psi_{\eta,q+1}(\tau)|^{1/2}. \tag{4.50}$$

A consistent estimator of $\psi_\varepsilon(\tau)$ can be obtained by replacing $\psi_{\eta,q+1}(\tau)$ with a consistent estimator in (4.50). To estimate $\psi_{n,q+1}(\tau)$, observe that

$$Y_{jt} - Y_{j,t-q-1} = (X_{jt} - X_{j,t-q-1})\beta + \varepsilon_{jt} - \varepsilon_{j,t-q-1}. \tag{4.51}$$

Therefore, $\psi_{\eta,q+1}$ is estimated consistently by the empirical characteristic function of the residuals from least-squares estimation of (4.51).

To estimate ψ_ξ, define $\eta_{jt,q} = \varepsilon_{jt} - \varepsilon_{j,t-q}$. Then

$$\eta_{j,t-q} = \xi_{jt} - (1-\alpha_q)\xi_{j,t-q} + \sum_{k=1}^{q-1} \alpha_k \xi_{j,t-k} - \sum_{k=1}^{q} \alpha_k \xi_{j,t-q-k} \, .$$

The characteristic function of $\eta_{jt,q}$ is

$$\psi_{\eta q}(\tau) = \psi_\xi(\tau)\psi_\xi[(1-\alpha_q)\tau]\psi_\xi(\alpha_q\tau)\prod_{k=1}^{q-1}\psi_\xi(\alpha_k\tau)^2 \, .$$

Therefore,

$$\frac{\psi_{\eta q}(\tau)}{[\psi_{\eta,q+1}(\tau)]^{1/2}} = \psi_\xi[(1-\alpha_q)\tau]\prod_{k=1}^{q-1}\psi_\xi(\alpha_k\tau)$$

and

$$r(\tau) \equiv \log\left\{\frac{\psi_{\eta q}(\tau)}{[\psi_{\eta,q+1}(\tau)]^{1/2}}\right\}$$

$$(4.52) \quad = \log\psi_\xi[(1-\alpha_q)\tau] + \sum_{k=1}^{q-1}\log\psi_\xi(\alpha_k\tau) \, .$$

Let $r_n(\tau)$ be a consistent estimator of $r(\tau)$, and let α_{nk} be a consistent estimator of α_k.. Then $\psi_\xi(\tau)$ can be estimated by carrying out the nonparametric mean regression of $r_n(\tau)$ on τ in a way that imposes the structure of the right-hand side of (4.52) after replacing α_k with α_{nk}. One way of doing this is through a power series approximation:

$$\log\psi_{n\xi}(\tau) = \sum_{j=1}^{K_n} a_j\tau^j \, ,$$

where $\psi_{n\xi}(\tau)$ is the estimator of $\psi_\xi(\tau)$, a_j ($j = 1, \ldots, K_n$) are constant coefficients, and $K_n \to \infty$ as $n \to \infty$. The coefficients a_j can be estimated by applying ordinary least squares to

$$r_n(\tau) = \sum_{j=1}^{K_n} a_j \left[(1-\alpha_{nq})^j + \sum_{k=1}^{q-1} \alpha_{nk}^j \right] \tau^j .$$

To form $r_n(\tau)$, let $\psi_{n\eta,q+1}$ be the empirical characteristic function of the residuals from least-squares estimation of (4.51). Let $\psi_{n\eta,q}$ be the empirical characteristic function of the residuals from least-squares estimation of

$$Y_{jt} - Y_{j,t-q} = (X_{jt} - X_{j,t-q})\beta + \varepsilon_{jt} - \varepsilon_{j,t-q} .$$

Then $r(\tau)$ is estimated consistently by

$$r_n(\tau) \equiv \log\left\{ \frac{\psi_{n\eta q}(\tau)}{[\psi_{n\eta,q+1}(\tau)]^{1/2}} \right\}$$

4.3.4 Asymmetrically Distributed ε's

This section shows how f_ε can be estimated when it is not assumed to be symmetrical about 0. It is assumed throughout this section that ε_{jt} is independently and identically distributed across both j and t. Section 4.3.5 shows how estimation can be carried out without assuming either symmetry of f_ε or serial independence of the ε_{jt}. Assume that $E(\varepsilon_{jt}) = 0$ for all j and t. This centering condition is automatic when ε_{jt} is symmetrically distributed but must be made explicitly when symmetry is not assumed.

Let ψ_ε and ψ_ε^*, respectively, denote the characteristic function of ε and its complex conjugate. Note that ψ_ε and ψ_ε^* are not equal if f_ε is not symmetrical about 0. If $\psi_{n\varepsilon}(\tau)$ is a consistent estimator of $\psi_\varepsilon(\tau)$ uniformly over $|\tau| \leq 1/\lambda_{n\varepsilon}$, f_ε can be estimated consistently by replacing $|\psi_{n\eta}(\tau)|^{1/2}$ with $\psi_{n\varepsilon}(\tau)$ in (4.21). To estimate $\psi_\varepsilon,(\tau)$, write

$$\psi_\varepsilon(\tau) = |\psi_\varepsilon(\tau)| \exp[i\omega(\tau)] ,$$

where $\omega(\tau)$ is the phase or argument of the complex-variable $\psi_\varepsilon(\tau)$. Observe that $\omega(0) = 0$ because $\psi_\varepsilon(0) = 1$ for any characteristic function. In addition, $\omega'(0) = 0$ because $E(\varepsilon_{jt}) = 0$ for all j and t. As in the case of symmetrical f_ε, $|\psi_{n\eta}(\tau)|^{1/2}$ is a consistent estimator of $|\psi_\varepsilon(\tau)|$. Therefore, the only remaining task is to construct an estimator of $\omega(\tau)$.

To do this, define $\eta_{jt} = \varepsilon_{jt} - \varepsilon_{j1}$. Define η_{njt} as in (4.18). Let $\Psi_\eta(\tau_2, \ldots, \tau_T)$ denote the joint characteristic function of $(\eta_{\bullet 2}, \ldots, \eta_{\bullet T})$, and let $\Psi_{n\eta}(\tau_2, \ldots, \tau_T)$ denote its sample analog:

$$\Psi_{n\eta}(\tau_2,\ldots,\tau_T) = \frac{1}{n}\sum_{j=1}^{n}\exp[i(\tau_2(\varepsilon_{j2}-\varepsilon_{j1})+\ldots+\tau_T(\varepsilon_{jt}-\varepsilon_{j1})].$$

Now because ε_{j1}, ε_{j2}, ..., ε_{jT} are independent random variables for each j,

$$\Psi_\eta(\tau_2,\ldots,\tau_T) = E\exp\{i[\tau_2(\varepsilon_{j2}-\varepsilon_{j1})+\ldots+\tau_T(\varepsilon_{jT}-\varepsilon_{j1})]\}$$

$$= E\exp[[i(\tau_2\varepsilon_{j2}+\ldots+\tau_T\varepsilon_{jT})]E\exp[-i(\tau_2+\ldots+\tau_T)\varepsilon_{j1}].$$

Therefore

$$\Psi_\eta(\tau,\tau,\ldots,\tau) = [\psi_\varepsilon(\tau)]^{T-1}\psi_\varepsilon[-(T-1)\tau]$$

and

$$\arg\Psi_\eta(\tau,\tau,\ldots,\tau) = (T-1)\omega(\tau) - \omega[(T-1)\tau].$$

Moreover,

$$\Psi_\eta(\tau,\tau,\ldots,\tau) = \left|\Psi_\eta(\tau,\tau,\ldots,\tau)\right|\exp(i\{(T-1)\omega(\tau)-\omega[(T-1)\tau]\}).$$

It follows that if the logarithm is treated as a function of a complex variable, then

$$\text{(4.53)} \qquad \operatorname{Im}\left\{\log\left[\frac{\Psi_\eta(\tau,\tau,\ldots,\tau)}{\left|\Psi_\eta(\tau,\tau,\ldots,\tau)\right|}\right]\right\} = (T-1)\omega(\tau) - \omega[(T-1)\tau].$$

A consistent estimator of $\omega(\tau)$ can be obtained by carrying out the nonparametric regression of $\operatorname{Im}\{\log[\Psi_{n\eta}(\tau, \tau, \ldots, \tau)/|\Psi_{n\eta}(\tau, \tau, \ldots, \tau)|]\}$ on τ in a way that imposes the structure of the right-hand side of (4.53). One way to do this is through a power-series approximation:

$$\omega_n(\tau) = \sum_{j=2}^{K_n} a_j\tau^j,$$

where $K_n \to \infty$ as $n \to \infty$ and the coefficients a_j are found by applying ordinary least squares to

$$\text{Im}\left\{\log\left[\frac{\Psi_{n\eta}(\tau,\tau,\ldots,\tau)}{\left|\Psi_{n\eta}(\tau,\tau,\ldots,\tau)\right|}\right]\right\} = \sum_{j=2}^{K_n} a_j [(T-1)-(T-1)^j]\tau^j .$$

4.3.5 Serially Correlated and Asymmetrically Distributed ε's

This section explains how f_ε and f_ξ can be estimated when ε_{jt} follows either an AR(1) or an MA(1) process for each j and it is not assumed that f_ε is symmetrical around 0. Assume that $T \geq 3$. To begin, suppose that ε_{jt} follows an AR(1) process. Then the arguments made in Section 4.3.1 for the AR(1) case show that

$$\psi_\varepsilon(\tau) = \frac{\psi_\eta[\tau/(1-\alpha)]}{\psi_\xi^*[\tau/(1-\alpha)]},$$

where ψ_ξ^* denotes the complex conjugate of ψ_ξ. ψ_η can be estimated using the methods described in Section (4.3.1). To estimate ψ_ξ, define $\zeta_{jt} = \xi_{jt} - \xi_{j1}$ ($t = 2, \ldots, T$). An estimable expression for ζ_{jt} can be obtained from (4.40) by observing that

$$\xi_{jt} - \xi_{j1} = (\xi_{jt} - \xi_{j,t-1}) + (\xi_{j,t-1} - \xi_{j,t-2}) + \ldots + (\xi_{j2} - \xi_{j1}) .$$

Therefore, ζ_{jt} can be estimated by summing residuals from least-squares estimation of (4.40). ψ_ξ can now be estimated by methods identical to those of Section 4.3.2 after replacing η_{jt} with ζ_{jt} and ε_{jt} with ξ_{jt}. Given an estimator of ψ_ξ, f_ε and f_ξ can be estimated by using the methods described in Section 4.3.1.

Now suppose that ε_{jt} follows an MA(1) process for each j. Define ξ_{jt} and η_{jt1} and η_{jt2} as in Section 4.3.1. Then the arguments made in Section 4.3.1 for the case of MA(1) ε_{jt} imply that

$$\left|\psi_\xi(\tau)\right| = \frac{\left|\psi_{\eta 1}[\tau/(1-\alpha)]\right|}{\left|\psi_{\eta 2}[\tau/(1-\alpha)]\right|^{1/2}} .$$

Therefore, $\psi_{\eta 1}$, $\psi_{\eta 2}$, and $|\psi_\xi|$ can be estimated using the methods of Section 4.3.1. Moreover,

$$\left|\psi_\varepsilon(\tau)\right| = \left|\psi_\xi(\tau)\right|\left|\psi_\xi(\alpha\tau)\right| ,$$

so $|\psi_\varepsilon|$ can be estimated once $|\psi_\xi|$ has been estimated. It remains to obtain estimators of the arguments or phases of ψ_ξ and ψ_ε. But $\psi_\varepsilon(\tau) = \psi_\xi(\tau)\psi_\varepsilon(\alpha\tau)$. Therefore,

$$\arg[\psi_\varepsilon(\tau)] = \arg[\psi_\xi(\tau)] + \arg[\psi_\xi(\alpha\tau)],$$

so it suffices to find an estimator of $\arg[\psi_\xi(\tau)]$.

To estimate $\arg[\psi_\xi(\tau)] \equiv \omega(\tau)$, observe that

$$\frac{\psi_{\eta 1}(\tau)}{|\psi_{\eta 2}(\tau)|^{1/2}|\psi_\xi[(1-\alpha)\tau]|} = \exp(i\{\omega(\tau) - \omega[(1-\alpha)\tau] - \omega(\alpha\tau)\}).$$

Therefore

$$\vartheta(\tau) \equiv \mathrm{Im}\left(\log\left\{\frac{\psi_{\eta 1}(\tau)}{|\psi_{\eta 2}(\tau)|^{1/2}|\psi_\xi[(1-\alpha)\tau]|}\right\}\right)$$

$$= \omega(\tau) - \omega[(1-\alpha)\tau] - \omega(\alpha\tau). \tag{4.54}$$

Let $\vartheta_n(\tau)$ be the quantity obtained from $\vartheta(\tau)$ by replacing α, ψ_η, and $|\psi_\xi|$ with consistent estimators. Then $\omega(\tau)$ can be estimated by carrying out a nonparametric regression of $\vartheta_n(\tau)$ on τ in a way that imposes the structure of the right-hand side of (4.54) after replacing α with the consistent estimator α_n. As in Sections 4.3.1 and 4.3.2, this can be done through a power-series approximation. To do this, write

$$\omega_n(\tau) = \sum_{j=1}^{K_n} a_j \tau^j,$$

where, as before, a_j ($j = 1, \ldots, n$) are constant coefficients and $K_n \to \infty$ as $n \to \infty$. Then the coefficients a_j can be estimated by applying least-squares to

$$\vartheta_n(\tau) = \sum_{j=1}^{K_n} [1 - \alpha_n^j - (1-\alpha_n)^j] a_j \tau^j.$$

4.4 An Empirical Example

This section presents an empirical example that illustrates the use of the methods described in Section 4.2. The example is taken from Horowitz and Markatou (1996) and consists of estimating indicators of the earnings mobility of individuals. Specifically, consider an individual whose earnings are 100(1 - α) percent of median earnings of individuals with the same age and education, where $\alpha = 0.10$ or 0.20. The example consists of using model (4.1) to estimate the probability that the individual's earnings never exceed 100(1 + α) percent of the median in any of the subsequent 2, 4, 6, 8, or 10 years. This corresponds to estimating $P(\theta \,|y_1, y^*, x)$, where $\theta = 3, 5, 7, 9$, or 11. The variable x specifies age and education; and y_1 and y^*, respectively, are 100(1 - α) and 100(1 + α) percent of median earnings conditional on x.

The estimation sample consists of 1643 white, male, full-time workers, aged 21-60 years and with at least 6 years of formal education, sampled randomly from the matched March 1986 and 1987 Current Population Survey. Each individual is included in the sample for each year, so the data form a panel of length $T = 2$.

The potential usefulness of nonparametric estimators such as those described in this chapter can be illustrated by a graphical analysis of the distribution of ε. Consider the residuals from estimation differenced model, η_{njt}, that are given by (4.18). If ε is normally distributed, these residuals are also normally distributed up to random sampling error. Therefore, normality of ε can be tested by testing for normality of the residuals. Let F_n denote the empirical distribution of these residuals, and let Φ denote the cumulative normal distribution function. If the residuals are normally distributed up to random sampling error, a plot of $\Phi^{-1}[F_n(v)]$ against v will consist of scatter around a straight line. The left-hand panel of Figure 4.2, which is taken from Horowitz and Markatou (1996), shows this plot. It is clear that ε is not normally distributed; the tails of its distribution are thicker than those of the normal distribution.

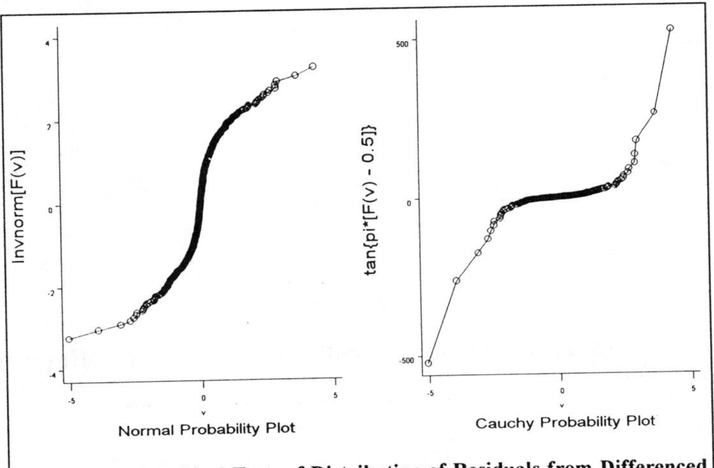

Figure 4.2: Graphical Tests of Distribution of Residuals from Differenced Model. Source: Horowitz and Markatou (1996).

Since the distribution of ε appears to be thick-tailed, one might consider approximating it with a Cauchy distribution. If ε is Cauchy distributed, the least-squares estimator of β in the differenced model is not consistent, but the least-absolute-deviations (LAD) estimator is. The residuals from the differenced model estimated by LAD will by Cauchy distributed up to random sampling error. A plot of $\tan\{\pi[F_n(v) - 0.5]\}$ against v will consist of scatter around a straight line. This is because $\tan\{\pi[F_n(\bullet) - 0.5]\}$ is the inverse of the cumulative standard Cauchy distribution function. The right-hand panel of Figure 4.2 shows the resulting plot. It is clear that ε is not Cauchy distributed; the tails of its distribution are too thin.

Of course, the fact that ε has neither the normal nor the Cauchy distribution does not rule out the possibility that it does belong to some other simple, parametric family of distributions that might be found through a careful specification search. However, a parametric model that is found through a specification search amounts to an informal nonparametric estimator whose statistical properties are unknown. In contrast, the formal nonparametric estimators described in this chapter have known properties.

A plot analogous to Figure 4.2 cannot be made for U because U is not observed. However, an informal graphical test of normality of U can be obtained by plotting $\log[\psi_{nU}(\tau)]$ against $-\tau^2$. If U is normally distributed, then $\psi_U(\tau) = \exp(-0.5\sigma_U^2\tau^2)$, so the plot will consist of scatter around a straight line. Figure 4.3 shows the plot. It suggests that any departure of the distribution of U from normality is mild. In the data used here, only ε has a distribution that is distinctly non-normal.

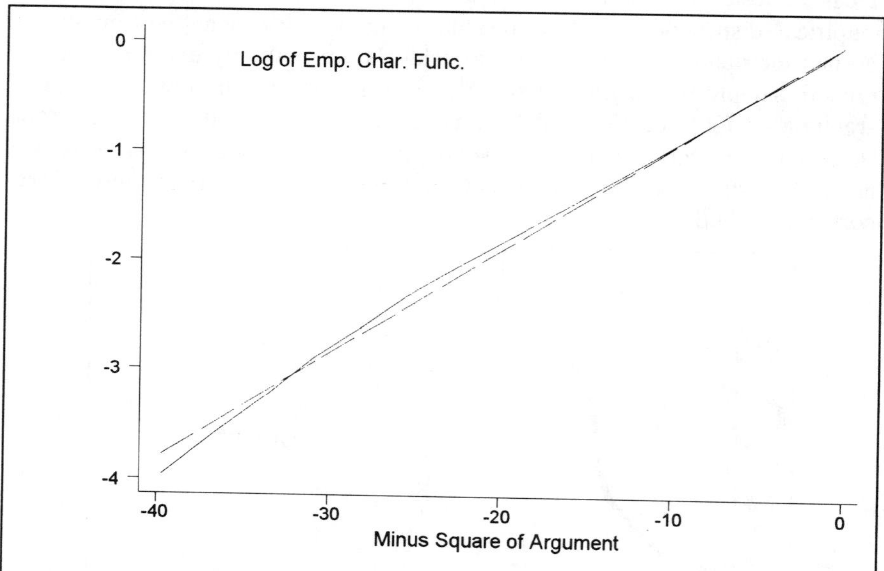

Figure 4.3: Graphical Test of Normality of *U*. Source: Horowitz and Markatou (1996).

$P(\theta|y_1,y^*,x)$ was estimated using the bias-corrected nonparametric estimators of f_ε and f_U described in this chapter. $P(\theta|y_1, y^*, x)$ was also estimated under the assumption that ε and U are both normally distributed. The assumption of normality is frequently made in applications though it is erroneous for ε in this example, as Figure 4.2 shows.

Table 4.2 shows the estimates. Depending on the values of α and θ, the estimates obtained from the nonparametric estimators of f_ε and f_U are 15-95 percent higher than those obtained by assuming that ε and U are normally distributed. Thus, the assumption that ε is normally distributed leads to a substantial overestimation of the probability that an individual with low earnings will become a high earner in the future. This finding is consistent with the results of the Monte Carlo experiments discussed in Section 4.2.4, which showed that the probabilities of transitions from low to high values of Y are overestimated if ε has a thick-tailed distribution but is assumed to be normally distributed.

Table 4.2: Semiparametric and Parametric Estimates of $P(\theta|y_1, y^*, x)$

α	θ	Estimate of P Based on Nonparametric Density Estimates	Estimate of P Based on Normality Assumption
0.10	3	0.71	0.62
	5	0.59	0.45
	7	0.51	0.34
	9	0.44	0.28
	11	0.39	0.23
0.20	3	0.97	0.73
	5	0.89	0.58
	7	0.82	0.49
	9	0.77	0.42
	11	0.72	0.37

Source: Horowitz and Markatou (1996)

Chapter 5
Transformation Models

This chapter is concerned with estimating models of the form

$$(5.1) \qquad T(Y) = X\beta + U ,$$

where T is a strictly increasing function, Y is an observed dependent variable, X is an observed random vector, β is a vector of constant parameters that is conformable with X, and U is an unobserved random variable that is independent of X. T is assumed to be strictly increasing to insure that (5.1) uniquely determines Y as a function of X and U. In applied econometrics, models of the form (5.1) are used frequently for the analysis of duration data and estimation of hedonic price functions. Familiar versions of (5.1) include the proportional hazards model, the accelerated failure time model, and the Box-Cox (1964) regression model.

Let F denote the CDF of U. The statistical problem addressed in this chapter is estimating β, T, and F when T and F are not both known. If T and F are known or known up to finite-dimensional parameters, then β and any parameters of T and F can be estimated by the method of maximum likelihood. In this chapter, it is assumed that at least one of the functions T and F does not belong to a known, finite-dimensional parametric family. Three cases are discussed. In Section 5.1, T is parametric (that is, known up to a finite-dimensional parameter) but F is not. In Section 5.2, F is parametric but T is not, and in Section 5.3, T and F are both nonparametric. Section 5.4 discusses the problem of predicting Y conditional on X from estimates of β, T, and F. Section 5.5 presents an empirical illustration of some of the methods that are discussed in Sections 5.1-5.4.

This chapter does not discuss estimation of (5.1) when Y is censored. Censoring occurs when one observes not Y but $\min(Y, C)$, where C is a variable that may be either fixed or random. Censoring often arises in the analysis of duration data, which is one of the main applications of transformation models. If Y is the duration of an event, censoring occurs if data acquisition terminates before all the events under observation have terminated. Most of the estimators described in this chapter can be adapted for use with censored data. Methods for doing this are given in references cited in Sections 5.1-5.3.

5.1 Estimation with Parametric T and Nonparametric F

In this section it is assumed that T is known up to a finite-dimensional parameter, α. In this case, (5.1) has the form

$$(5.2) \qquad T(Y, \alpha) = X\beta + U,$$

where $T(\bullet,\bullet)$ is a known function and U is independent of X. It can be assumed without loss of generality that $E(U) = 0$, and this will be done except as otherwise noted. The discussion in this section begins by presenting some important forms of T. Then methods for estimating α, β, and F are discussed.

One well-known form of T is the Box-Cox (1964) transformation:

$$(5.3) \qquad T(y, a) = \begin{cases} \dfrac{y^a - 1}{a} & \text{if } a \neq 0 \\ \log y & \text{if } a = 0. \end{cases}$$

This transformation is widely used but has the disadvantage it requires Y in (5.2) to be non-negative unless a is a non-zero integer. Moreover, $T(y, a)$ is bounded from below (above) if $a > 0$ ($a < 0$) unless a is an odd integer or 0. Therefore, the Box-Cox transformation cannot be applied to models in which the dependent variable can be negative or the distribution of U has unbounded support (e.g., U is normally distributed).

Bickel and Doksum (1981) proposed overcoming this problem by assuming that $a > 0$ and replacing (5.3) with

$$(5.4) \qquad T(y, a) = \frac{|y|^a \operatorname{sgn}(y) - 1}{a}.$$

MacKinnon and Magee (1990) have pointed out that (5.4) also can have disadvantages in applications. In particular, $T(y,a)$ does not have a finite limit as $a \to 0$ if y is negative and is very steep near $y = 0$ if a is small.

MacKinnon and Magee (1990) proposed the family of transformations

$$T(y, a) = \frac{H(ay)}{a},$$

where H is an increasing function that satisfies $H(0) = 0$, $H'(0) = 1$, and $H''(0) \neq 0$. Of course, this family also has certain undesirable properties. It does not nest $\log y$ if y is positive, and it does not permit transformations that are skew-symmetric about 0. Other forms of T have been proposed by John (1949) and

John and Draper (1980). Every parametric family of transformations, however, has characteristics that make it unsuitable for certain applications. This is one of the motivations for using models in which T is nonparametric. Such models are discussed in Sections 5.2 and 5.3.

Regardless of the form of T, the inferential problem in (5.2) is to estimate α, β, and F from observations of (Y, X). Denote the estimation data by $\{Y_i, X_i: i = 1, ..., n\}$, and assume that they are a random sample from the joint distribution of (Y, X). Let a_n and b_n denote the estimators of α and β, respectively.

Given consistent estimators if α and β, estimation of F is straightforward. Suppose that a_n and b_n are $n^{1/2}$-consistent. Suppose, also, that F is differentiable and that T is a differentiable function of both of its arguments. Then, it is not difficult to show that F is estimated $n^{1/2}$-consistently by the empirical distribution function of the residuals $T(Y_i, a_n) - X_i b_n$ $(i = 1, ..., n)$.

Now consider the problem of estimating α and β. One estimation method that may seem attractive is nonlinear least squares. That is, a_n and b_n solve

$$\underset{a,b}{\text{minimize:}} \quad S_n(a,b) = \frac{1}{n}\sum_{i=1}^{n}[T(Y_i, a) - X_i b]^2 .$$

Unfortunately, however, the resulting estimators are inconsistent for α and β.

To see why, observe that the first-order conditions for minimizing S_n are

$$\text{(5.5)} \quad \frac{\partial S_n(a,b)}{\partial a} = \frac{2}{n}\sum_{i=1}^{n}\frac{\partial T(Y_i, a)}{\partial a}[T(Y_i, a) - X_i b] = 0$$

and

$$\text{(5.6)} \quad \frac{\partial S_n(a,b)}{\partial b} = -\frac{2}{n}\sum_{i=1}^{n} X_i'[T(Y_i, a) - X_i b] = 0 .$$

Suppose that T and $\partial T(Y,a)/\partial a$ are continuous functions of a. Suppose, also, that a_n and b_n converge almost surely to α and β. Then it follows from the uniform strong law of large numbers (Jennrich 1969) that

$$\frac{\partial S_n(a_n, b_n)}{\partial a} \to E\frac{\partial S_n(\alpha,\beta)}{\partial a}$$

and

$$\frac{\partial S_n(a_n, b_n)}{\partial b} \to E\frac{\partial S_n(\alpha,\beta)}{\partial b}$$

almost surely as $n \to \infty$. Therefore,

$$E\frac{\partial S_n(\alpha,\beta)}{\partial a} = E\frac{\partial S_n(\alpha,\beta)}{\partial b} = 0.$$

if $a_n \to \alpha$ and $b_n \to \beta$ almost surely. But $T(Y_i, \alpha) - X_i\beta = U_i$. Therefore

$$E\frac{\partial S_n(\alpha,\beta)}{\partial a} = E\left[\frac{\partial T(Y,\alpha)}{\partial a}U\right]$$

and

$$E\frac{\partial S_n(\alpha,\beta)}{\partial a} = E(X'U).$$

Now, $E(X'U) = 0$ because $E(U) = 0$ and U and X are independent. But $E\{[\partial T(Y,\alpha)/\partial a]U\} \neq 0$ in general because Y is correlated with U. Therefore, the assumption that a_n and b_n are consistent for α and β leads to a contradiction. It follows that the nonlinear least-squares estimators of α and β are inconsistent.

Examination of (5.5) and (5.6) shows that the nonlinear least-squares estimator is an instrumental variables estimator in which the instruments are X and an estimate of $\partial T(Y,\alpha)/\partial a$. As is well known, consistent instrumental-variables estimation requires that the instruments be uncorrelated with U. Viewed from this perspective, the cause of inconsistency of the nonlinear least-squares estimator is that $\partial T(Y,\alpha)/\partial a$ is correlated with U and, therefore, is not a valid instrument. This observation suggests that a consistent estimator can be obtained by replacing $\partial T(Y,\alpha)/\partial a$ in (5.6) with a valid instrument.

Let W be a column vector of valid instruments. Validity requires that $E(WU) = 0$ and $\dim(W) \geq \dim(\beta) + 1$. Powers, cross-products, and other nonlinear functions of components of X can be used to form W. The choice of instruments is discussed further in Section 5.1.1. Given W, α and β can be estimated by the generalized method of moments (GMM) based on the population moment condition

$$(5.7) \qquad E\{W[T(Y,\alpha) - X\beta]\} = 0,$$

provided that this equation uniquely determines α and β. The estimator solves

$$(5.8) \qquad \underset{a,b}{\text{minimize:}} \quad G_n(a,b)'\Omega_n G_n(a,b),$$

where

$$G_n(a,b) = \frac{1}{n}\sum_{i=1}^{n} W_i[T(Y_i,a) - X_i b],$$

and Ω_n is a positive-definite, possibly stochastic matrix. One possible choice of Ω_n is $(WW')^{-1}$, in which case (5.8) gives the nonlinear, two-stage least squares (NL2SLS) estimator of α and β (Amemiya 1985). Amemiya and Powell (1981) discuss the use of NL2SLS for estimation of (5.2) when T is the Box-Cox transformation.

It is also possible to obtain an estimator of $(\alpha, \beta')'$ that is asymptotically efficient for the specified instruments, W, even if U is not independent of X, provided that $E(U|X = x) = 0$. To do this, set Ω_n equal to the following estimator of $\Omega_0 \equiv \{EWW'[T(Y,\alpha) - X\beta]^2\}^{-1}$:

$$(5.9) \quad \Omega_{n0} = \left\{\frac{1}{n}\sum_{i=1}^{n} W_i W_i'[T(Y_i,\tilde{a}_n) - X_i\tilde{b}_n]^2\right\}^{-1},$$

where $(\tilde{a}_n, \tilde{b}_n')'$ is a preliminary consistent estimator of $(\alpha, \beta')'$ (possibly the NL2SLS estimator).

If U in (5.2) is independent of X, then NL2SLS is the asymptotically efficient estimator based on (5.7), and the second estimation stage is not needed. However, NS2SLS does not fully exploit the implications of the independence assumption. When independence holds, further improvements in asymptotic efficiency may be possible by augmenting (5.7) with moment conditions based on higher-order moments of U. See Newey (1993) for further discussion of this possibility.

The asymptotic distributional properties of GMM estimators were derived by Hansen (1982). Define $\theta = (\alpha, \beta')'$ and $\theta_n = (a_n, b_n')'$. Hansen (1982) showed that under mild regularity conditions, θ_n is a consistent estimator of θ. Moreover,

$$(5.10) \quad n^{1/2}(\theta_n - \theta) \to^d N(0,V),$$

where

$$(5.11) \quad V = (D'\Omega D)^{-1} D'\Omega\Omega_0^{-1}\Omega D(D'\Omega D)^{-1},$$

$$D = E\frac{\partial}{\partial\theta}W[T(Y,\alpha) - X\beta],$$

and

$$\Omega = \operatorname*{plim}_{n\to\infty} \Omega_n .$$

V can be estimated consistently by replacing D in (5.11) by

$$D_n = \frac{\partial G_n(\alpha, \beta)}{\partial \theta},$$

Ω with Ω_n, and Ω_0 with Ω_{n0}. Thus, (5.10) and (5.11) make it possible to carry out inference about θ in sufficiently large samples.

5.1.1 Choosing the Instruments

It is clear from (5.11) that the covariance matrix of the asymptotic distribution of $n^{1/2}(\theta_n - \theta)$ depends on the choice of instruments, W. This suggests choosing instruments that maximize the asymptotic efficiency of the estimator.

The problem of optimal choice of instruments has been investigated by Newey (1990, 1993). Define $\sigma^2(x) = \text{Var}(U|X = x)$ and

$$\Delta(x) = E\left\{\frac{\partial}{\partial \theta}[T(Y, \alpha) - X\beta | X = x]\right\}. \tag{5.12}$$

Then the optimal instruments for estimating θ are

$$W_{opt}(X) = C\sigma^{-2}(X)\Delta(X),$$

where C is any nonsingular matrix. If U is independent of X, then $\sigma^2(x)$ is a constant and can be dropped from the expression for W_{opt}. If W_{opt} were known, the asymptotic efficiency of an estimator of $(\alpha, \beta')'$ would be maximized by solving (5.8) with $W_i = W_{opt}(X_i)$ and $\Omega_n = \Omega_{n0}$.

Of course, this estimator is not feasible because α, β, $\sigma^2(x)$, and the conditional expectation in (5.12) are unknown. However, Newey (1990, 1992) has shown that the estimator remains asymptotically efficient if the unknown quantities are replaced by suitable consistent estimators. Specifically, α and β can be replaced with consistent but inefficient estimators, such as $\tilde{a}_n$ and $\tilde{b}_n$. The conditional expectation in (5.12) can be replaced by a nonparametric regression estimator of the conditional mean of $(\partial/\partial\theta)[T(Y, \tilde{a}_n) - X\tilde{b}_n]$, and $\sigma^2(x)$ can be replaced with a nonparametric regression estimator of the conditional variance of $[T(Y, \tilde{a}_n) - X\tilde{b}_n]$. Newey suggests using nearest-neighbor or series nonparametric mean-regression estimators to avoid technical difficulties that are associated with the random denominators of kernel estimators. However, kernel estimators can undoubtedly be made to work as well. See Newey (1990, 1993) for detailed discussions of the nearest-neighbor and series estimation approaches.

5.1.2 The Weibull Hazard Model with Unobserved Heterogeneity

The Weibull hazard model with unobserved heterogeneity is a special case of (5.2) that is important in the analysis of duration data and unusually difficult to estimate. To describe this model, define $P(\bullet|x)$ to be the CDF of Y conditional on $X = x$, and define $p(\bullet|x)$ to be the conditional density of Y. The conditional hazard of Y at the point (y, x) is defined as the density of Y at y conditional on $Y \geq y$ and $X = x$. Formally, the conditional hazard function, $\lambda(y|x)$ is

$$(5.13) \qquad \lambda(y|x) = \frac{p(y|x)}{1 - P(y|x)}.$$

Let F and f, respectively, denote the CDF and density function of U. Then $P(y|x) = F[T(y,\alpha) - x\beta]$, and

$$(5.14) \qquad \lambda(y|x) = \frac{dT(y,\alpha)}{dy} \frac{f[T(y,\alpha) - x\beta]}{1 - F[T(y,\alpha) - x\beta]}.$$

The Weibull hazard model is obtained by setting $T(y,\alpha) = \log(y^{\alpha})$ for some $\alpha > 0$ and assuming that U has the extreme value distribution whose CDF is $F(u) = 1 - \exp(-e^{u})$. Substituting these functions into (5.14) yields the conditional hazard function

$$(5.15) \qquad \lambda(y|x) = \alpha y^{\alpha-1} e^{-x\beta}.$$

Substitution of $T(y,\alpha) = \log(y^{\alpha})$ into (5.2) gives the equivalent model

$$(5.16) \qquad \alpha \log Y = X\beta + U,$$

where $F(u) = 1 - \exp(-e^{u})$. In this model, $E(U)$ is a non-zero constant, rather than zero as assumed in (5.2). However, this change of the location of U can be absorbed in the coefficient of an intercept component of X and, therefore, has no substantive significance.

The Weibull hazard model with unobserved heterogeneity is obtained from (5.15) and (5.16) by assuming that in addition to the observed covariates, X, there is an unobserved covariate, also called unobserved heterogeneity or frailty. For example, if Y is the duration of a spell of unemployment, then the unobserved covariate might represent personal characteristics that affect an individual's attractiveness to employers but are unobserved by the analyst. The unobserved covariate enters the model as an unobserved random variable, V. Thus, in the Weibull hazard model with unobserved heterogeneity, (5.15) and (5.16) become

$$(5.17) \qquad \lambda(y|x) = \alpha y^{\alpha-1} e^{-x\beta-V}$$

and

$$(5.18) \qquad \alpha \log Y = X\beta + V + U .$$

In this section, it is assumed that V is independent of X and U but that its distribution is otherwise unknown. If the distribution of V were known up to a finite-dimensional parameter, then (5.17) and (5.18) would be a fully parametric model that could be estimated by maximum likelihood. See, for example, Lancaster (1979).

When (5.18) holds, the moment condition (5.7) becomes

$$E[W(\alpha \log Y - X\beta)] = 0 .$$

Equivalently,

$$(5.19) \qquad \alpha E[W(\log Y - X\gamma)] = 0 .$$

where $\gamma = \beta/\alpha$. Equation (5.19) reveals why the Weibull hazard model with unobserved heterogeneity is difficult to estimate. Equation (5.19) holds for any value of α. Therefore, (5.19) and (5.7) do not identify α and cannot be used to form estimators of α. Equation (5.19) does identify γ, however. In fact, γ can be estimated consistently by applying ordinary least-squares to

$$(5.20) \qquad \log Y = X\gamma + \nu ,$$

where ν is a random variable whose mean is zero.

Although (5.7) and (5.19) do not identify α, this parameter is, nonetheless, identified if $e^{-V} < \infty$. The distribution of V is also identified. Elbers and Ridder (1982), Heckman and Singer (1984a), and Ridder (1990) provide detailed discussions of identification in the proportional hazards model with unobserved heterogeneity. The fact that α and the distribution of V are identified suggests that they and $\beta = \alpha\gamma$ should be estimable, though not by GMM based on (5.7) or (5.19).

Before discussing methods for estimating α, β, and the distribution of V, it is useful to consider how rapidly such estimators can be expected to converge. The parameter $\gamma = \beta/\alpha$ can be estimated with a $n^{-1/2}$ rate of convergence by applying ordinary least squares to (5.20). Ishwaran (1997) has shown that the fastest possible rate of convergence of an estimator of α and, therefore, of β is $n^{-d/(2d+1)}$ if $d > 0$ is the largest number such that $Ee^{\pm dV} < \infty$. Thus, α and β cannot be estimated with $n^{-1/2}$ rates of convergence.

Estimating the distribution of V is a problem in deconvolution. The fastest possible rate of convergence of an estimator of the CDF or density of V is a power of $(\log n)^{-1}$. To see why, let Γ denote the CDF of V, and define $W = \alpha \log Y - X\beta$. Then $W = V + U$, and the distribution of W is the convolution of the distribution of U, which is known, and the unknown distribution of V. If α and β were known, then each observation of Y and X would produce an observation of W. The problem of estimating the distribution of V from observations of W would then be a deconvolution problem of the type discussed in Section 4.1. As was explained in Section 4.1, the rate of convergence of an estimator of Γ is determined by the thickness of the tails of the characteristic function of U. It can be shown that in the Weibull hazard model, where $F(u) = 1 - \exp(-e^u)$, the tails of the characteristic function of U decrease exponentially fast. Therefore, for reasons explained in Section 4.1, an estimator of Γ can have at most a logarithmic rate of convergence. Of course, α and β are not known in applications, but lack of knowledge of these parameters cannot accelerate the convergence of an estimator of Γ. Thus, the rate-of-convergence result for known α and β also holds when these parameters are unknown.

Methods for estimating α, β, and Γ will now be discussed. Heckman and Singer (1984b) suggested estimating α, β, and the distribution of V simultaneously by nonparametric maximum likelihood estimation. To set up this approach, observe that by (5.18), the density of Y conditional on $X = x$ and $V = v$ is

$$(5.21) \qquad p(y|x,v,\alpha,\beta) = \alpha y^{\alpha-1} e^{-x\beta-v} \exp\left(-y^{\alpha} e^{-x\beta-v}\right).$$

Therefore, the density of Y conditional on $X = x$ is

$$(5.22) \qquad p(y|x,\alpha,\beta,\Gamma) = \int \alpha y^{\alpha-1} e^{-x\beta-v} \exp\left(-y^{\alpha} e^{-x\beta-v}\right) d\Gamma(v)$$

Now let $\{Y_i, X_i: i = 1, ..., n\}$ be a random sample of the joint distribution of (Y,X). The log-likelihood of the sample at parameter values a, b, and G is

$$(5.23) \qquad \log L_n(a,b,G) = \sum_{i=1}^{n} p(Y_i|X_i,a,b,G)$$

The estimator of Heckman and Singer (1984b) is obtained by maximizing $\log L_n$ over a, b, and G.

The maximum likelihood estimator of Heckman and Singer (1984b) is unconventional because it entails maximizing over an infinite-dimensional parameter (the function G) as well as the finite-dimensional parameters a and b. Kiefer and Wolfowitz (1956) have given conditions under which maximum likelihood estimators are consistent in models with infinitely many parameters. Heckman and Singer (1984b) show that their estimator satisfies the conditions of Kiefer and Wolfowitz (1956) and, therefore, is consistent for α, β, and Γ. The rate of convergence and asymptotic distribution of the Heckman-Singer estimator are unknown.

The result of Heckman and Singer (1984b) is stated formally in the following theorem.

Theorem 5.1: Let (a_n, b_n, G_n) maximize $\log L_n$ in (5.23). As $n \to \infty$, $a_n \to \alpha$ almost surely, $b_n \to \beta$ almost surely, and $G_n(z) \to \Gamma(z)$ almost surely for each z that is a continuity point of Γ if the following conditions hold:

(a) $\{Y_i, X_i: i = 1, ..., n\}$ is a random sample of (Y, X). The conditional density of Y is given by (5.22).

(b) α and the components of β are contained in bounded open intervals.

(c) The components of X have finite first absolute moments. The support of the distribution of X is not contained in any proper linear subspace of $\Re^{\ell}$, where ℓ is the number of components of X.

(d) The distribution of V is non-defective. That is, it does not have mass points at $-\infty$ or ∞.

(e) $Ee^{-V} < \infty$. ■

This completes the discussion of the Heckman-Singer estimator.

Honoré (1990) has developed an estimator of α that has known asymptotic properties. Honoré considers, first, a model in which there are no covariates. The density of Y in this model, $p(\bullet)$ can be obtained from (5.22) by setting $\beta = 0$. This gives

$$p(y) = \int \alpha y^{\alpha-1} e^{-v} \exp\left(-y^{\alpha} e^{-v}\right) d\Gamma(v)$$

The CDF of Y can be obtained by integrating $p(y)$ and is

$$P(y) = \int \exp\left(-y^{\alpha} e^{-v}\right) d\Gamma(v) .$$

Honoré's estimator is based on the observation that

$$(5.24) \qquad \alpha = \lim_{y \to 0} \frac{\log\{-\log[1 - P(y)]\}}{\log y},$$

as can be verified by using l'Hospital's rule. The estimator is obtained by using order statistics of Y to construct a sample analog of the right-hand side of (5.24). The details of the construction are intricate and will not be given here. To state the result, let $Y_{m:n}$ denote the m'th order statistic of $Y_1, \ldots, Y_n$. Define $m_1 = n^{1-\delta_1}$ and $m_2 = n^{1-\delta_2}$, where $0 < \delta_2 < \delta_1 < 1$. Define

$$\rho = 1 - \frac{1}{2} \frac{n^{-\delta_1} - n^{-\delta_2}}{(\delta_1 - \delta_2)\log n}.$$

Honoré's estimator of α is

$$(5.25) \qquad a_n = -\frac{\rho(\delta_1 - \delta_2)\log n}{\log Y_{m_1:n} - \log Y_{m_2:n}}.$$

Honoré shows that if $P(e^{-V} > 0) > 0$, $Ee^{-2V} < \infty$ and $\delta_1 + 2\delta_2 > 1$, then the rate of convergence of a_n to α is $(\log n)^{-1} n^{-(1-\delta_1)/2}$. This rate is always slower than $n^{-1/3}$, though it can be made close to $n^{-1/3}$ by choosing δ_1 and δ_2 appropriately. Honoré also shows that the centered and normalized form of a_n is asymptotically normally distributed. Specifically

$$(5.26) \qquad (a_n \sigma)^{-1}(a_n - \alpha) \to^d N(0,1),$$

where

$$(5.27) \qquad \sigma^2 = \left(\frac{1}{(\delta_1 - \delta_2)\log n}\right)^2 \frac{n^{\delta_1} - n^{\delta_2}}{n}.$$

Equations (5.26) and (5.27) make it possible to carry out statistical inference about α.

Ishwaran (1996) derived an estimator of α whose rate of convergence is $n^{-1/3}$, which is the fastest possible rate when e^{-V} is assumed to have only two moments. Ishwaran (1996) does not give the asymptotic distribution of his estimator.

In models with covariates, α can be estimated by ignoring the covariates (that is, treating them as if they were unobserved) and applying (5.25). Given estimates a_n and γ_n of α and γ, β can be estimated by $b_n = a_n\gamma_n$. Honoré (1990) does not discuss estimation of the distribution of V. This can be done, however, by applying the deconvolution methods of Chapter 4 to the estimates of W_i (i = 1, ..., n) that are given by $W_{ni} = a_n \log Y_i - X_i b_n$.

5.2 Estimation with Nonparametric T and Parametric F

This section treats the version of (5.1) in which T is an unknown increasing function but the distribution of U is known or known up to a finite-dimensional parameter. To begin, observe that (5.1) is unchanged if any constant, c is added to the right-hand side and T is replaced by the function T^* that is defined by $T^*(y) = T(y) + c$. Therefore, a location normalization is needed to make identification possible. Here, location normalization will be accomplished by omitting an intercept component from X. Thus, all components of X are non-degenerate random variables, and all components of β are slope coefficients.

5.2.1 The Proportional Hazards Model

The proportional hazards model of Cox (1972) is widely used for the analysis of duration data. This model is most frequently formulated in terms of the hazard function of the non-negative random variable Y conditional on covariates X. This form of the model is

$$\text{(5.28)} \qquad \lambda(y|x) = \lambda_0(y)e^{-x\beta} .$$

In this model, $\lambda(y|x)$ is the hazard that $Y = y$ conditional on $X = x$. The non-negative function λ_0 is called the *baseline hazard function*. The essential characteristic of (5.28) that distinguishes it from other models is that $\lambda(y|x)$ is specified to be the product of a function of y alone and a function of x alone.

To put (5.28) into the form (5.1), integrate (5.13) with respect to y to obtain

$$\int_0^y \lambda(\tau|x)d\tau = -\log[1 - P(y|x)]$$

so that

$$\text{(5.29)} \qquad P(y|x) = 1 - \exp\left[-\int_0^y \lambda(\tau|x)d\tau\right].$$

Define the integrated baseline hazard function, Λ_0 by

$$\Lambda_0(y) = \int_0^y \lambda_0(\tau)d\tau .$$

Then substitution of (5.28) into (5.29) gives

$$(5.30) \qquad P(y|x) = 1 - \exp\left[-\Lambda_0(y)e^{-x\beta}\right]$$

under the proportional hazards model. It is easily shown that (5.30) is also obtained by assuming that Y satisfies

$$(5.31) \qquad \log \Lambda_0(Y) = X\beta + U ,$$

where U has the CDF $F(u) = 1 - \exp(-e^u)$. Therefore, (5.31) is the version of (5.1) that yields the Cox (1972) proportional hazards model.

Equation (5.31) contains two unknown quantities, β and Λ_0. Both can be estimated from data. Consider, first, estimation of β. Define the *risk set* at y as $R(y) = \{i: \ Y_i \geq y\}$. Cox (1972) proposed estimating β by maximizing the *partial likelihood function* that is defined by

$$(5.32) \qquad L_{np}(b) = \prod_{i=1}^{n} \left[\frac{\exp(-X_i b)}{\sum_{j \in R(Y_i)} \exp(-X_j b)} \right]$$

Equivalently, β can be estimated by maximizing the logarithm of the partial likelihood function. This yields the *partial likelihood estimator* of β.

To understand the partial likelihood estimator intuitively, it is useful to think of Y as the duration of some event such as a spell of unemployment. Y then has units of time. Now consider the risk set at time Y_i. Suppose it is known that among the events in this set, exactly one terminates at time Y_i. The probability density that event j terminates at time Y_i given that it is in the risk set is simply the conditional hazard, $\lambda(Y_i|X_j)$. Since the probability that two events terminate simultaneously is zero, the probability density that some event in the risk set terminates at time Y_i is

$$\sum_{j \in R(Y_i)} \lambda(Y_i | X_j) .$$

Therefore, the probability density that the terminating event is i is

$$\frac{\lambda(Y_i | X_i)}{\sum_{j \in R(Y_i)} \lambda(Y_i | X_j)}$$

or, by using (5.28),

$$\frac{\exp(-X_i\beta)}{\sum_{j\in R(Y_i)} \exp(-X_j\beta)}.$$

Thus, the partial likelihood function at $b = \beta$ can be interpreted as the probability density of the observed sequence of terminations of events conditional on the observed termination times and risk sets.

The asymptotic properties of the partial likelihood estimator have been investigated by Tsiatis (1981) and by Andersen and Gill (1982). Let b_n denote the partial likelihood estimator. Tsiatis (1981) has shown that under regularity conditions, b_n converges to β almost surely as $n \to \infty$. Thus, b_n is strongly consistent for β. Tsiatis (1981) has also shown that

$$(5.33) \qquad n^{1/2}(b_n - \beta) \to^d N(0,V)$$

where

$$(5.34) \qquad V = -\left[E\frac{1}{n}\frac{\partial \log L_{np}(\beta)}{\partial b \partial b'}\right]^{-1}.$$

V is estimated consistently by

$$V_n = -\left[\frac{1}{n}\frac{\partial \log L_{np}(b_n)}{\partial b \partial b'}\right]^{-1},$$

thereby making statistical inference about β possible. The proof of asymptotic normality of $n^{1/2}(b_n - \beta)$ is an application of standard Taylor series methods of asymptotic distribution theory. See Amemiya (1985) for a discussion of these methods.

The asymptotic normality result may be stated formally as follows:

Theorem 5.2: Let $\{Y_i, X_i: i = 1, ..., n\}$ be a random sample of (Y, X). Assume that the conditional hazard function of Y is given by (5.28) and that $E[X'Xe^{-2X\beta}] < \infty$. Then the equation

$$\frac{\partial \log L_{np}(b)}{\partial b} = 0$$

has a sequence of solutions $\{b_n\}$ that converges almost surely to β as $n \to \infty$. Equations (5.33) and (5.34) hold for this sequence. ■

The integrated baseline hazard function, Λ_0, can also be estimated. Let P_X denote the CDF of X. Define $Q(y) = P(Y > y)$. Then

$$Q(y) = 1 - \int P(y|x)dP_X(x)$$

$$= \int \exp\left[-\Lambda_0(y)e^{-x\beta}\right]dP_X(x)$$

Also define

$$H(y) = \int e^{-x\beta} \exp\left[-\Lambda_0(y)e^{-x\beta}\right]dP_X(x).$$

Then some algebra shows that

$$\Lambda_0(y) = -\int_0^y \frac{1}{H(\tau)}dQ(\tau). \tag{5.35}$$

Tsiatis (1981) proposed estimating Λ_0 by replacing the right-hand side of (5.35) with a sample analog. The resulting estimator is

$$\Lambda_{n0}(y) = \sum_{\{i: Y_i \le y\}} \frac{1}{\sum_{j \in R(Y_i)} \exp(X_j b_n)}, \tag{5.36}$$

where b_n is the partial likelihood estimator of β. Under regularity conditions, $\Lambda_{n0}(y)$ is a consistent estimator of $\Lambda_0(y)$, and $n^{1/2}[\Lambda_{n0}(y) - \Lambda_0(y)]$ is asymptotically normally distributed with mean zero and a variance that can be estimated consistently. See Tsiatis (1981) for the details of the variance and its estimator. In fact, Tsiatis (1981) proves the much stronger result that $n^{1/2}[\Lambda_{n0}(y) - \Lambda_0(y)]$ converges weakly to a mean-zero Gaussian process. A mean-zero Gaussian process is a type of random function, W, with the property (among others) that the random variables $W(y_1)$, $W(y_2)$, ..., $W(y_k)$ for any finite k are multivariate normally distributed with mean zero. See Billingsley (1968) and Pollard (1984) for formal definitions of a Gaussian process and weak convergence and for discussions of other mathematical details that are needed to make the weak convergence results stated here precise.

In many applications, the baseline hazard function, λ_0, is of greater interest than the integrated baseline hazard function Λ_0. However, λ_0 cannot be estimated by differentiating Λ_{n0} in (5.36) because Λ_{n0} is a step function. A method for estimating λ_0 is discussed in Section 5.2.4.

5.2.2 The Proportional Hazards Model with Unobserved Heterogeneity

In the proportional hazards model with unobserved heterogeneity, an unobserved covariate V enters the conditional hazard function in addition to the observed covariates X. The hazard function of the dependent variable Y conditional on $X = x$ and $V = v$ is

$$(5.37) \qquad \lambda(y|x,v) = \lambda_0(y)e^{-x\beta - v}$$

Model (5.37) is equivalent to

$$\log \Lambda_0(Y) = X\beta + V + U ,$$

where Λ_0 is the integrated baseline hazard function and U has the CDF $F(u) = 1 - \exp(-e^u)$. The Weibull hazard model with unobserved heterogeneity that was discussed in Section 5.1.2 is the special case of (5.37) that is obtained by assuming that $\lambda_0(y) = y^\alpha$ for some $\alpha > 0$. As was discussed in Section 5.1.2, V is often interpreted as representing unobserved attributes of individuals that affect the duration of the event of interest. In Section 5.1.2 it was assumed that λ_0 is known up to the parameter α but that the distribution of V is unknown. In this section the assumptions are reversed. It is assumed that λ_0 is unknown and that the distribution of V is known up to a finite-dimensional parameter.

Most approaches to estimating model (5.37) under these assumptions assume that e^{-V} has a gamma distribution with mean 1 and unknown variance θ. Thus, if $Z = e^{-V}$, the probability density function of Z is

$$f_Z(z) = \frac{\theta^{-1/\theta}}{\Gamma(1/\theta)} z^{1/\theta - 1} e^{-z/\theta},$$

where Γ is the gamma function. Hougaard (1984) and Clayton and Cusick (1985) provide early discussions of this model and its use. See Hougaard (1986) and Lam and Kuk (1997) for discussions of other potentially useful parametric families of distributions of Z. The assumption $E(e^{-V}) = 1$ is a normalization that is needed for identification and has no substantive consequences. This can be seen by noting that if $E(e^{-V}) = \mu \neq 1$, then an algebraically equivalent model can be obtained by replacing V with $V' = V - \log\mu$ and λ_0 with $\lambda_0' = \mu\lambda_0$. In the replacement model, $E(e^{-V'}) = 0$.

In addition, to assuming that e^{-V} is gamma distributed, most research on model (5.37) has assumed that there are no observed covariates X. To minimize the complexity of the discussion, only this case will be treated formally here. The extension to models with covariates will be discussed informally. See Parner (1997a, 1997b) for a formal treatment of models with covariates.

The inferential problem to be solved is estimating θ and Λ_0, without assuming that Λ_0 belongs to a known, finite-dimensional parametric family of functions. This problem can be solved by forming a maximum likelihood estimator of θ and Λ_0.

To form the likelihood function, observe that by arguments identical to those used to derive (5.30), the CDF of Y conditional on $Z = z$ is $P(Y \le y|Z = z) = 1 - \exp[-z\Lambda_0(y)]$. Therefore, the conditional density of Y is $f(y|z) = z\lambda_0(y)\exp[-z\Lambda_0(y)]$, and the joint density of Y and Z is $f(y|z)f_Z(z)$ or

$$(5.38) \qquad f_{YZ}(y,z) = \frac{\theta^{-1/\theta}}{\Gamma(1/\theta)} z^{1/\theta} e^{-z/\theta} \lambda_0(y) \exp[-z\Lambda_0(y)].$$

The marginal density of Y is obtained by integrating $f_{YZ}(y, z)$ with respect to z. This yields

$$(5.39) \qquad f_Y(y) = \frac{\lambda_0(y)}{[1+\theta\Lambda_0(y)]^{1+1/\theta}}.$$

Equation (5.39) suggests the possibility of estimating θ and Λ_0 by maximizing the likelihood function obtained from f_Y. However, the resulting estimator of Λ_0 is a step function with jumps at the observed values of Y. To form a likelihood function that accommodates step functions, let $\varepsilon > 0$ be an arbitrarily small number, and let $\Delta\Lambda_0(y)$ denote the change in Λ_0 over the interval y to $y + \varepsilon$. Then, the probability that Y is observed to be in the interval y to $y + \varepsilon$ is

$$f_Y(y)\varepsilon = \frac{\Delta\Lambda_0(y)}{[1+\theta\Lambda_0(y)]^{1+1/\theta}}.$$

It follows that for observations $\{Y_i: i = 1, ..., n\}$ that are a random sample of Y, the log-likelihood function is

$$(5.40) \qquad \log L(t, A) = \sum_{i=1}^{n} \left\{ \log[\Delta A(Y_i)] - \left(1+\frac{1}{t}\right) \log[1+tA(Y_i)] \right\},$$

where $A(y)$ is the generic parameter for $\Lambda_0(y)$ and $\Delta A(y)$ is the jump in A at the point y. The maximum likelihood estimator of θ and Λ_0 is obtained by maximizing the right-hand side of (5.39) over t and A.

Maximization of $\log L$ in (5.40) entails computing the estimates of $n + 1$ parameters, namely the estimate of θ and the estimates of the n jumps of $\Delta\Lambda_0$.

Nielsen *et al.* (1992) and Petersen, Andersen, and Gill (1996) have pointed out that the computations can be simplified greatly through the use of the EM algorithm of Dempster, Laird, and Rubin (1977). To implement this algorithm, suppose for the moment that Z were observable and Z_i were the i'th observation. Then Λ_0 could be estimated by (5.36) with Z_j in place of $X_j b_n$. The EM algorithm replaces the unobservable Z_j with an estimator, Z_{ni}. The estimator of $\Lambda_0(y)$ is then

$$(5.41) \qquad \Lambda_{n0}(y) = \sum_{\{i: Y_i \leq y\}} \frac{1}{\sum_{j \in R(Y_i)} \exp(Z_{nj})}$$

Given this estimator of Λ_0, θ can be estimated by maximizing $\log L$ over t while holding A fixed at Λ_{n0}. Computing the estimate of θ this way is a one-dimensional optimization that is much easier to carry out than the $(n + 1)$-dimensional optimization required to maximize $\log L$ directly. Z_j is estimated by an estimator of $E(Z_j|Y_1, \ldots, Y_n)$. It is not difficult to show that

$$E(Z_i | Y_1, \ldots, Y_n) = \frac{1+\theta}{1+\theta\Lambda_0(Y_i)}.$$

This is estimated by

$$(5.42) \quad Z_{ni} = \frac{1+\theta_n}{1+\theta_n \Lambda_{n0}(Y_i)},$$

where θ_n is an estimate of θ. The EM algorithm consists of iterating (5.41), (5.42), and the one-dimensional maximization of $\log L$ until convergence. Specifically, it consists of repeating the following steps until convergence is achieved (Petersen, Andersen and Gill 1996):

Step 1: Begin with an initial estimate of Λ_0.

Step 2: Let $\Lambda_{n0}^{(k)}$ denote the estimate of Λ_0 at iteration k of the algorithm. Obtain $\theta_n^{(k)}$, the iteration k estimate of θ, by maximizing $\log L$ with respect to t while holding A fixed at $\Lambda_{n0}^{(k)}$.

Step 3 (E step): Compute the iteration k estimate of Z_i $(i = 1, \ldots, n)$ by applying (5.42) with $\theta_n^{(k)}$ and $\Lambda_{n0}^{(k)}$ in place of θ_n and Λ_{n0}.

Step 4 (M step): Compute $\Lambda_{n0}^{(k+1)}$ by applying (5.40) with $Z_{nj}^{(k)}$ in place of Z_{nj}. Return to step 2.

Murphy (1994, 1995) has derived the asymptotic properties of the maximum likelihood estimators of θ and Λ_0. Let θ_n and Λ_{n0} denote the estimators. Murphy (1994) has shown that as $n \to \infty$, $\theta_n \to \theta$ almost surely and $\Lambda_{n0}(y) \to \Lambda_0(y)$ almost surely uniformly over y in bounded intervals. Murphy (1995) has shown that $n^{1/2}(\theta_n - \theta)$ is asymptotically normal with mean zero and that $n^{1/2}(\Lambda_{n0} - \Lambda_0)$ converges weakly to a Gaussian process whose mean is zero. The latter result implies $n^{1/2}[\Lambda_{n0}(y) - \Lambda_0(y)]$ is asymptotically normal with mean zero for any given y. See Murphy (1995) for expressions for the variance of the asymptotic distribution of $n^{1/2}(\theta_n - \theta)$ and the covariance function of the limiting Gaussian process for $n^{1/2}(\Lambda_{n0} - \Lambda_0)$

The extension of the maximum likelihood estimator to a model with covariates will now be outlined. The model described here is a special case of one considered by Parner (1997a, 1997b) in which V is permitted to be correlated across observations. See Parner (1997a, 1997b) for a formal discussion of the asymptotic properties of the maximum likelihood estimator.

When (5.37) holds with covariates X and gamma-distributed $Z \equiv e^{-V}$, arguments similar to those leading to (5.38) show that the joint density of Y and Z conditional on $X = x$ is

$$f_{YZ}(y,z|x) = \frac{\theta^{-1/\theta}}{\Gamma(1/\theta)} z^{1/\theta} e^{-z/\theta} \lambda_0(y)e^{-x\beta} \exp[-z\Lambda_0(y)e^{-x\beta}]$$

Integration over z gives the density of Y conditional on $X = x$ as

$$(5.43) \qquad f_Y(y|x) = \frac{\lambda_0(y)e^{-x\beta}}{[1+\theta\Lambda_0(y)e^{-x\beta}]^{1+1/\theta}}.$$

Therefore, the log-likelihood function is

$$(5.44) \qquad \log L(t,b,A) = \sum_{i=1}^{n} \left\{ \log[\Delta A(Y_i)e^{-X_i b}] - \left(1+\frac{1}{t}\right) \log[1+tA(Y_i)e^{-X_i b}] \right\}.$$

Maximum likelihood estimation can now be carried out by using the following modified version of the previously described EM algorithm:

Step 1: Begin with an initial estimate of Λ_0.

Step 2′: Let $\Lambda_{n0}^{(k)}$ denote the estimate of Λ_0 at iteration k of the algorithm. Obtain $\theta_n^{(k)}$ and $b_n^{(k)}$, the iteration k estimates of θ and β, by maximizing $\log L$ in (5.44) with respect to t and b while holding A fixed at $\Lambda_{n0}^{(k)}$.

Step 3′ (E step): Compute the iteration k estimate of Z_i ($i = 1, ..., n$) by applying (5.42) with $\theta_n^{(k)}$ and $\Lambda_{n0}^{(k)}(Y_i)\exp[-X_i b_n^{(k)}]$ in place of θ_n and $\Lambda_{n0}(Y_i)$.

Step 4′ (M step): Compute $\Lambda_{n0}^{(k+1)}$ by applying (5.40) with $Z_{nj}^{(k)}\exp[-X_j b_n^{(k)}]$ in place of Z_{nj}. Return to step 2.

5.2.3 The Case of Discrete Observations of Y

The complexity of the estimators of θ, β, and Λ_0 in Section 5.2.2 is due to the need to estimate the function Λ_0, which takes values over a continuum and, therefore, is an infinite-dimensional parameter. The estimation problem can be simplified greatly by assuming that one observes not Y but only which of finitely many intervals of the real line contains Y. This approach was taken by Meyer (1990), who analyzed duration data that were rounded to integer numbers of weeks. See, also, Prentice and Gloeckler (1978). When the interval containing Y is observed but Y is not, $\Lambda_0(y)$ is identified only if y is a boundary point of one of the intervals. Therefore, it is necessary to estimate Λ_0 at only finitely many points. This reduces estimation of θ, β, and Λ_0 to a finite-dimensional problem. The properties of maximum likelihood estimators of these parameters can be obtained from the standard theory of maximum likelihood estimation of finite-dimensional parametric models.

To see how estimation can be carried out when Y is observed only in intervals, let $\{y_j : j = 0, 1, ..., K\}$ denote the boundaries of the intervals. Define $y_0 = 0$, and assume that $y_K < \infty$. It follows from (5.43) that for $1 \leq j \leq K$

$$P(y_{j-1} < Y \leq y_j | X = x) = \int_{y_{j-1}}^{y_j} f_Y(\xi | x) d\xi$$

$$= \left[1 + \theta\Lambda_0(y_{j-1})e^{-x\beta}\right]^{-1/\theta} - \left[1 + \theta\Lambda_0(y_j)e^{-x\beta}\right]^{-1/\theta}$$

and

$$P(Y > y_K | X = x) = \left[1 + \theta\Lambda_0(y_K)e^{-x\beta}\right]^{-1/\theta}.$$

Therefore, the log-likelihood of a random sample of (Y, X) is

$$\log L(t,b,A) = \sum_{i=1}^{n} \sum_{j=1}^{K} I(y_{j-1} < Y_i \le y_j) \log\left[\left(1+tA_{j-1}e^{X_i b}\right)^{1/t} - \left(1+tA_j e^{X_i b}\right)^{1/t}\right]$$

$$+ \sum_{i=1}^{n} I(Y_i > y_K) \log\left(1+tA_K e^{X_i b}\right)^{1/t},$$

where $A = (A_0, ..., A_K)$ is the generic parameter vector for $[\Lambda_0(y_0), ..., \Lambda_0(y_K)]$, and $A_0 = 0$. The unknown parameters of the model are θ, β, and $\Lambda_0(y_j)$ $(1 \le j \le K)$. These can be estimated by maximizing $\log L(t,b,A)$ over t, b, and A. The total number of parameters to be estimated is $1 + \dim(X) + K$, regardless of the size of n. Therefore, the estimation problem is finite-dimensional and the theory of maximum likelihood estimation of finite-dimensional parametric models applies to the resulting estimators.

5.2.4 Estimating λ_0

There are many applications of the proportional hazards model with or without unobserved heterogeneity in which the baseline hazard function λ_0 is of interest. For example, if Y is the duration of a spell of unemployment, λ_0 indicates whether the hazard of terminating unemployment increases, decreases or varies in a more complicated way as the duration of unemployment increases. This information can be useful for testing substantive explanations of the process of finding new employment. An increasing hazard of terminating unemployment, for example, might be expected if an unemployed individual searches for a job with increasing intensity as the duration of unemployment increases. This section explains how to estimate λ_0.

In the model discussed in Section 5.1.2, $\lambda_0(y) = y^{\alpha}$, so an estimator of λ_0 can be formed by replacing α with its estimator, a_n, in the expression for λ_0. The situation is more complicated in the models of Sections 5.2.1 and 5.2.2, where Λ_0 is nonparametric. Since $\lambda_0(y) = d\Lambda_0(y)/dy$, one might consider estimating λ_0 by $d\Lambda_{n0}(y)/dy$, where Λ_{n0} is the estimator of Λ_0 in (5.36). This procedure does not work, however, because Λ_{n0} is a step function

A similar problem arises in using an empirical distribution function, F_n, to estimate a probability density function f. A density estimator cannot be formed by differentiating F_n because F_n is a step function. In density estimation, this problem is solved by smoothing the empirical distribution function to make it differentiable. The same technique can be applied to Λ_{n0}.

Let K be a kernel function of a scalar argument, possibly a probability density function. Let $\{h_n\}$ be a sequence of positive numbers (bandwidths) that converges to zero as $n \to \infty$. The kernel estimator of $\lambda_0(y)$ is

$$(5.44)\qquad \lambda_{n0}(y) = \frac{1}{h_n}\int K\left(\frac{y-\xi}{h_n}\right) d\Lambda_{n0}(\xi)\,.$$

Note the similarity between (5.44) and a kernel density estimator, which is what would be obtained if Λ_{n0} were replaced with F_n in the integral. The remainder of this section provides a heuristic argument showing that $\lambda_{n0}(y)$ has properties that are similar to those of a kernel density estimator. In particular, $\lambda_{n0}(y) \to^p \lambda_0(y)$ as $n \to \infty$ at a rate that is no faster than $n^{-2/5}$ if $y > 0$ and λ_0 is twice continuously differentiable in a neighborhood of y. Similar arguments can be used to show that a faster rate of convergence is possible if λ_0 has more than two derivatives and K is a higher-order kernel. As in nonparametric density estimation, however, a rate of convergence of $n^{-1/2}$ is not possible.

To begin the analysis of λ_{n0}, write it in the form

$$(5.45)\qquad \lambda_{n0}(y) = \lambda_0(y) + H_{n1}(y) + H_{n2}(y)\,,$$

where

$$H_{n1}(y) = \frac{1}{h_n}\int K\left(\frac{y-\xi}{h_n}\right) d\Lambda_0(\xi) - \lambda_0(y)$$

and

$$(5.46)\qquad H_{n2}(y) = \frac{1}{h_n}\int K\left(\frac{y-\xi}{h_n}\right) d[\Lambda_{n0}(\xi) - \Lambda_0(\xi)]\,.$$

Now consider H_{n1}. Observe that it can be written

$$(5.47)\qquad H_{n1}(y) = \frac{1}{h_n}\int K\left(\frac{y-\xi}{h_n}\right) \lambda_0(\xi) d\xi\,.$$

Assume that K is a continuously differentiable probability density function that is symmetrical about zero and whose support is [-1,1]. Making the change of variables $\zeta = (\xi - y)/h_n$ in the integral on the right-hand side of (5.47) then yields

$$H_{n1}(y) = \int_{-1}^{1} K(\zeta)\lambda_0(h_n\zeta + y)d\zeta\,.$$

Now expand $\lambda_0(h_n\zeta + y)$ in a Taylor series to obtain

$$H_{n1}(y) = \int_{-1}^{1} K(\zeta)[\lambda_0(y) + h_n\zeta\lambda_0'(y) + (1/2)h_n^2\zeta^2\lambda_0''(y) + o(h_n^2)]d\zeta - \lambda_0(y)$$

$$= (1/2)h_n^2\lambda_0''(y)A_K + o(h_n^2),$$

where

$$A_K = \int_{-1}^{1} \zeta^2 K(\zeta)d\zeta.$$

Substitution of this result into (5.45) yields

$$\text{(5.48)} \qquad \lambda_{n0}(y) = \lambda_0(y) + (1/2)h_n^2\lambda_0''(y)A_K + H_{n2}(y) + o(h_n^2).$$

Now consider $H_{n2}(y)$. Integrate by parts on the right-hand side of (5.46) to obtain

$$H_{n2}(y) = \frac{1}{h_n^2}\int [\Lambda_{n0}(\xi) - \Lambda_0(\xi)]K'\left(\frac{y-\xi}{h_n}\right)d\xi$$

$$= \frac{1}{n^{1/2}h_n^2}\int n^{1/2}[\Lambda_{n0}(\xi) - \Lambda_0(\xi)]K'\left(\frac{y-\xi}{h_n}\right)d\xi$$

As was discussed in Sections 5.2.1 and 5.2.2, $n^{1/2}(\Lambda_{n0} - \Lambda_0)$ converges weakly to a Gaussian process. Let W denote the limiting process. Given any finite, $y_1 > 0$ and $y_2 > 0$, define $V(y_1, y_2) = E[W(y_1)W(y_2)]$. $V(y_1, y_2)$ is the covariance of the jointly normally distributed random variables $W(y_1)$ and $W(y_2)$. Using the theory of empirical processes (see, e.g., Billingsley 1968 or Pollard 1984), it can be shown that

$$H_{n2}(y) = \hat{H}_{n2}(y) + o_p(n^{-1/2}),$$

where

$$\hat{H}_{n2}(y) = \frac{1}{n^{1/2}h_n^2}\int W(\xi)K'\left(\frac{y-\xi}{h_n}\right)d\xi$$

Moreover,

$$\hat{H}_{n2}(y) \sim N(0, \sigma_{ny}^2),$$

where for any finite $y > 0$

$$\sigma_{ny}^2 = E[\hat{H}_{n2}(y)^2]$$

$$= E\frac{1}{nh_n^4}\int W(\xi)W(\zeta)K'\left(\frac{y-\xi}{h_n}\right)K'\left(\frac{y-\zeta}{h_n}\right)d\xi d\zeta$$

$$= \frac{1}{nh_n^4}\int V(\xi,\zeta)K'\left(\frac{y-\xi}{h_n}\right)K'\left(\frac{y-\zeta}{h_n}\right)d\xi d\zeta.$$

By a change of variables in the integrals

$$\text{(5.49)} \qquad \sigma_{ny}^2 = \frac{1}{nh_n^2}\int_{-1}^{1} d\xi \int_{-1}^{1} d\zeta V(h_n\xi + y, h_n\zeta + y)K'(\xi)K'(\zeta).$$

Because V is a symmetrical function of its arguments, $\sigma_{ny}{}^2$ can also be written in the form

$$\sigma_{ny}^2 = \frac{2}{nh_n^2}\int_{-1}^{1} d\xi \int_{-1}^{\xi} d\zeta V(h_n\xi + y, h_n\zeta + y)K'(\xi)K'(\zeta).$$

It is not difficult to show that

$$\int_{-1}^{1} d\xi \int_{-1}^{\xi} d\zeta K'(\xi)K'(\zeta) = 0,$$

and

$$\int_{-1}^{1} d\xi \int_{-1}^{\xi} d\zeta \xi K'(\xi)K'(\zeta) = -\int_{-1}^{1} d\xi \int_{-1}^{\xi} d\zeta \xi K'(\xi)K'(\zeta) = \frac{1}{2}B_K,$$

where

$$B_K = \int_{-1}^{1} K(\xi)^2 d\xi .$$

Therefore, a Taylor series expansion of the integrand on the right-hand side of (5.49) yields

$$(5.50) \qquad \sigma_{ny}^2 = \frac{B_K}{nh_n} V_1(y, y) + o\left(\frac{1}{nh_n}\right),$$

where

$$V_1(y, y) = \lim_{\xi \to 0+} \frac{\partial V(\xi + y, y)}{\partial \xi} .$$

It follows from (5.48) and (5.50) that the asymptotic mean-square error of $\lambda_{n0}(y)$ is minimized when $h_n = cn^{-1/5}$ for some $c > 0$. It also follows that with this h_n, $\lambda_{n0}(y) - \lambda_0(y) = O_p(n^{-2/5})$ and

$$(nh_n)^{1/2}[\lambda_{n0}(y) - \lambda_0(y)] \to^d N(\mu_y, \sigma_y^2),$$

where $\mu_y = (1/2)c^{5/2}A_K\lambda_0''(y)$ and $\sigma_y^2 = B_K V_1(y,y)$. Thus, $n^{2/5}$-consistency of $\lambda_{n0}(y)$ and asymptotic normality of $(nh_n)^{1/2}[\lambda_{n0}(y) - \lambda_0(y)]$ are established.

Implementing (5.44) requires choosing the value of the c. One possible choice is the value of c that minimizes the asymptotic mean-square error of $\lambda_{n0}(y) - \lambda_0(y)$. When $h_n = cn^{-2/5}$, The asymptotic mean-square error is

$$AMSE(y) = \frac{\mu_y^2 + \sigma_y^2}{nh_n}$$

$$= \frac{c^4[A_K \lambda_0''(y)]^2}{4n^{4/5}} + \frac{B_K V_1(y, y)}{cn^{4/5}} .$$

$AMSE(y)$ is minimized by setting $c = c_{opt}$, where

$$c_{opt} = \left\{ \frac{B_K V_1(y, y)}{[A_K \lambda_0''(y)]^2} \right\}^{1/5} .$$

This value of c can be estimated by the plug-in method, which consists of replacing V_1 and λ_0'' in the expression for c_{opt} with estimators.

5.2.5 Other Models in which F Is Known

In this section, it is assumed that (5.1) holds with T unknown and F known. In contrast to Section 5.2.1, however, it is not assumed that $F(u) = 1 - \exp(-e^u)$. Instead, F can be any known distribution function that satisfies certain regularity conditions. The aim is to estimate the finite-dimensional parameter β and the transformation function T. The methods described in this section can be applied to the proportional hazards model of Section 5.2.1, although there is no reason to do so. They can also be applied to the *proportional odds* model (Pettit 1982, Bennett 1983a,b), which is obtained from (5.1) by assuming U to be logistically distributed. The methods described in this section cannot be applied to the proportional hazards model with unobserved heterogeneity described in Section 5.2.2 unless the variance parameter θ is known *a priori*.

The estimators described in this section are due to Cheng, Wei and Ying (1995, 1997). These estimators have the advantage that they are applicable to models with a variety of different F's. If interest centers on a particular F, it may be possible to take advantage of special features of this F to construct estimators that are asymptotically more efficient than those of Cheng, Wei, and Ying. For example, the partial likelihood estimator of β in the proportional hazards model is asymptotically efficient (Bickel, *et al.* 1993). Murphy, Rossini, and Van der Vaart (1997) show that a semiparametric maximum likelihood estimator of β in the proportional odds model is asymptotically efficient. However, an estimator that takes advantage of features of a specific F may not be consistent with other F's.

Consider, now, the problem of estimating β in (5.1) with a known F. Let (Y_i, X_i) and (Y_j, X_j) $(i \neq j)$ be two distinct, independent observations of (Y, X). Then it follows from (5.1) that

$$
\begin{aligned}
&E[I(Y_i > Y_j) | X_i = x_i, X_j = x_j] \\
(5.51) \qquad &= P[U_i - U_j > -(x_i - x_j)\beta \,|\, X_i = x_i, X_j = x_j].
\end{aligned}
$$

Let $G(z) = P(U_i - U_j > z)$ for any real z. Then

$$
G(z) = \int_{-\infty}^{\infty} [1 - F(u + z)] dF(u) .
$$

G is a known function because F is assumed known in this Section. Substituting G into (5.51) gives

$$
E[I(Y_i > Y_j) | X_i = x_i, X_j = x_j] = G[(-x_i - x_j)\beta] .
$$

Define $X_{ij} = X_i - X_j$. Then it follows that β satisfies the moment condition

$$(5.52) \qquad E\{w(X_{ij}\beta)X_{ij}[I(Y_i > Y_j) - G(-X_{ij}\beta)]\} = 0,$$

where w is a weight function. Cheng, Wei, and Ying (1995) propose estimating β by replacing the population moment condition (5.52) with the sample analog

$$(5.53) \qquad \sum_{i=1}^{n} \sum_{j=1}^{n} \{w(X_{ij}b)X_{ij}[I(Y_i > Y_j) - G(-X_{ij}b)]\} = 0.$$

The estimator of β, b_n, is the solution to (5.53). Equation (5.53) has a unique solution if $w(z) = 1$ for all z and the matrix $\Sigma_i\Sigma_j(X_{ij}'X_{ij})$ is positive definite. It also has a unique solution asymptotically if w is positive everywhere (Cheng, Wei, and Ying 1995). Moreover, b_n converges almost surely to β. This follows from uniqueness of the solution to (5.53) and almost sure convergence of the left-hand side of (5.53) to $E\{w(X_{ij}b)X_{ij}[I(Y_i > Y_j) - G(-X_{ij}b)]\}$.

The asymptotic distribution of $n^{1/2}(b_n - \beta)$ can be obtained by using standard Taylor-series methods of asymptotic distribution theory. To this end, define

$$H_n(b) = \frac{1}{n^2} \sum_{i=1}^{n} \sum_{j=1}^{n} \{w(X_{ij}b)X_{ij}[I(Y_i > Y_j) - G(-X_{ij}b)]\} = 0$$

Then $H_n(b_n) = 0$, and a Taylor-series approximation gives

$$(5.54) \qquad n^{1/2} H_n(\beta) + \frac{\partial H_n(b_n^*)}{\partial b} n^{1/2}(b_n - \beta) = 0,$$

where b_n^* is between b_n and β, and $\partial H_n/\partial b$ is the matrix whose (j,k) element is the partial derivative of the j'th component of H_n with respect to the k'th component of b. Cheng, Wei, and Ying (1995) show that $n^{1/2}H_n(\beta)$ is asymptotically normally distributed with mean zero and a covariance matrix equal to the probability limit of

$$(5.55) \qquad Q_n \equiv \frac{1}{n^3} \sum_{i=1}^{n} \sum_{j=1}^{n} \sum_{\substack{k=1 \\ k \neq j}}^{n} [w(X_{ij}b_n)e_{nij} - w(X_{ji}b_n)e_{nji}] \cdot [w(X_{ik}b_n)e_{nik} - w(X_{ki}b_n)e_{nki}]X_{ij}'X_{ik},$$

where $e_{nij} = I(Y_i > Y_j)[1 - G(X_{ij}\beta)]$. Cheng, Wei, and Ying (1995) also show that if G is everywhere differentiable, then $\partial H_n(b_n^*)/\partial b$ converges in probability to the probability limit of

$$(5.56) \qquad R_n = \frac{1}{n^2} \sum_{i=1}^{n} \sum_{j=1}^{n} w(X_{ij}b_n)G'(X_{ij}b_n)X_{ij}'X_{ij} .$$

Let Q and R be the probability limits of Q_n and R_n, respectively. Then it follows from (5.54)-(5.56) that $n^{1/2}(b_n - \beta) \rightarrow^d N(0, R^{-1}QR^{-1})$ if w is positive and G is differentiable. The covariance matrix $R^{-1}QR^{-1}$ is estimated consistently by $R_n^{-1}Q_nR_n^{-1}$. This result enables statistical inference about β to be carried out.

Now consider estimation of the transformation function T. This problem is addressed by Cheng, Wei, and Ying (1997). Equation (5.1) implies that for any real y and vector x that is conformable with X, $E[I(Y \leq y)|X = x] - F[T(y) - x\beta] = 0$. Cheng, Wei, and Ying (1997) propose estimating $T(y)$ by the solution to the sample analog of this equation. That is, the estimator $T_n(y)$ solves

$$\frac{1}{n} \sum_{i=1}^{n} \{I(Y_i \leq y) - F[t(y) - X_i b_n]\} = 0 ,$$

where b_n is the solution to (5.53). Cheng, Wei, and Ying (1997) show that if F is strictly increasing on its support, then $T_n(y)$ converges to $T(y)$ almost surely uniformly over any interval $[0,\tau]$ such that $P(Y > \tau) > 0$. Moreover, $n^{1/2}(T_n - T)$ converges to a mean-zero Gaussian process over this interval. The covariance function of this process is lengthy and is given in Cheng, Wei, and Ying (1997).

5.3 Estimation when Both T and F Are Nonparametric

This section discusses estimation of (5.1) when neither T nor F is assumed to belong to a known, finite-dimensional parametric family of functions. The aim is to develop $n^{1/2}$-consistent estimators of β, T, and F. The estimators described here are taken from Horowitz (1996b). Other estimators that are not $n^{1/2}$-consistent have been described by Breiman and Friedman (1985) and Hastie and Tibshirani (1990). Gørgens and Horowitz (1995) show how to estimate β, T, and F when observations of Y are censored.

When T and F are nonparametric, certain normalizations are needed to make identification of (5.1) possible. First, observe that (5.1) continues to hold if T is replaced by cT, β is replaced by $c\beta$, and U is replaced by cU for any positive constant c. Therefore, a scale normalization is needed to make identification possible. This will be done here by setting $|\beta_1| = 1$, where β_1 is the first component of β. Observe, also, that when T and F are nonparametric, (5.1) is a semiparametric single-index model. Therefore, as was discussed in Chapter 2, identification of β requires X to have at least one component whose distribution conditional on the others is continuous and whose β coefficient is non-zero. Assume without loss of generality that the components of X are ordered so that the first satisfies this condition.

It can also be seen that (5.1) is unchanged if T is replaced by $T + d$ and U is replaced by $U + d$ for any positive or negative constant d. Therefore, a location normalization is also needed to achieve identification when T and F are nonparametric. Location normalization will be carried out here by assuming that $T(y_0) = 0$ for some finite y_0 that satisfies conditions given in Section 5.3.2. With this location normalization, there is no centering assumption on U and no intercept term in X.

The location and scale normalizations in this section are not the same as those of Sections 5.1 and 5.2. When T or F is known up to a finite-dimensional parameter, as is the case in Sections 5.1 and 5.2, location and scale normalization are usually accomplished implicitly through the specification of T or F. This is not possible when T and F are nonparametric. Because of differences in normalization, estimates obtained using the methods described in this section cannot be compared with estimates obtained using the methods of Sections 5.1 and 5.2 without first making adjustments to give all estimates the same location and scale normalizations. To see how this can be done, let T_n, b_n, and F_n denote the estimates of T, β, and F obtained using the methods of this section. Let T_n^*, b_n^*, and F_n^* denote the estimates using another method. Let b_{n1}^* denote the first component of b_n^*. Then T_n^*, b_n^*, and F_n^* can be adjusted to have the location and scale normalization of T_n, b_n, and F_n by replacing $T_n^*(y)$ with $T_n^{**}(y) \equiv [T_n^*(y) - T_n^*(y_0)]/|b_{n1}^*|$, b_n^* with $b_n^{**} \equiv b_n^*/|b_{n1}^*|$, and $F_n^*(u)$ by $F_n^{**}(u) \equiv F_n^*[|b_{n1}^*|u + T_n^*(y_0)]$.

Now consider the problem of estimating T, β, and F with the normalizations $|\beta_1| = 1$ and $T(y_0) = 0$. Because (5.1) with nonparametric T and F is a semiparametric single-index model, β can be estimated using the methods described in Chapter 2. Let b_n denote the estimator of β. Estimators of T and F are derived in Section 5.3.1. The statistical properties of the estimators of T and F are described in Section 5.3.2.

5.3.1 Derivation of Estimators of T and F

This section derives nonparametric estimators of T and F. Consider, first, estimation of T. The estimator of Horowitz (1996b) is derived in two steps. In the first step, T is expressed as a functional of the population distribution of (Y, X). In the second step, unknown population quantities in this functional are replaced with sample analogs.

To take the first step, define $Z = X\beta$. Z is a continuous random variable because X is assumed to have at least one continuously distributed component with a non-zero β coefficient. Let $G(\bullet|z)$ be the CDF of Y conditional on $Z = z$. Assume that G is differentiable with respect to both of its arguments. Define $G_y(y|z) = \partial G(y|z)/\partial y$ and $G_z(y|z) = \partial G(y|z)/\partial z$. Equation (5.1) implies that

$$G(y|z) = F[T(y) - z].$$

Therefore,

$$G_y(y|z) = T'(y)F'[T(y) - z],$$

$$G_z(y|z) = -F'[T(y) - z],$$

and for any (y, z) such that $G_z(y|z) \neq 0$,

$$T'(y) = -\frac{G_y(y|z)}{G_z(y|z)}.$$

It follows that

$$T(y) = -\int_{y_0}^{y} \frac{G_y(v|z)}{G_z(v|z)} dv \tag{5.57}$$

for any z such that the denominator of the integrand is non-zero over the range of integration.

Now let $w(\bullet)$ be a scalar-valued function on $\Re$ with compact support S_w such that (a) the denominator of the integrand in (5.57) is non-zero for all $z \in S_w$ and $v \in [y_0, y]$, and (b)

$$\int_{S_w} w(z)dz = 1. \tag{5.58}$$

Then

$$T(y) = -\int_{y_0}^{y} \int_{S_w} w(z) \frac{G_y(v|z)}{G_z(v|z)} dz dv. \tag{5.59}$$

Equation (5.59) is the desired expression for T as a functional of the population distribution of (Y, X). This completes the first step of the derivation of the estimator of Horowitz (1996b).

The second step of the derivation consists of replacing the unknown quantities on the right-hand side of (5.59) with consistent estimators. The unknown parameter β is replaced by b_n. The unknown function $G(y|z)$ is replaced by a nonparametric kernel estimator, $G_n(y|z)$ of the CDF of Y conditional on $Xb_n = z$. G_z in (5.59) is replaced by $G_{nz} = \partial G_n/\partial z$. G_y is replaced by a kernel estimator, G_{ny}, of the probability density function of Y conditional on $Xb_n = z$. The resulting estimator of $T(y)$ is

$$(5.60) \qquad T_n(y) = -\int_{y_0}^{y} \int_{S_w} w(z) \frac{G_{ny}(v|z)}{G_{nz}(v|z)} dz dv .$$

To specify the estimators G_{ny} and G_{nz}, let $\{Y_i, X_i: i = 1, ..., n\}$ denote a random sample of (Y, X). Define $Z_{ni} = X_i b_n$. Let K_Y and K_Z be kernel functions of a scalar argument. These are required to satisfy conditions that are stated in Section 5.3.2. Among other things, K_Z must be a higher-order kernel. Let $\{h_{ny}\}$ and $\{h_{nz}\}$ be sequences of bandwidths that converge to zero as $n \to \infty$. Estimate $p(\bullet)$, the probability density function of Z, by

$$p_n(z) = \frac{1}{nh_{nz}} \sum_{i=1}^{n} K_Z\left(\frac{Z_{ni} - z}{h_{nz}}\right)$$

The estimator of $G(y|z)$ is

$$(5.61) \qquad G_n(y|z) = \frac{1}{nh_{nz} p_n(z)} \sum_{i=1}^{n} I(Y_i \le y) K_Z\left(\frac{Z_{ni} - z}{h_{nz}}\right).$$

The estimator of $G_{nz}(y|z)$ is $\partial G_n(y|z)/\partial z$. $G_y(y|z)$ is the probability density function of Y conditional on $Z = z$. It cannot be estimated by $\partial G_n(y|z)/\partial y z$ because $G_n(y|z)$ is a step function of y. Instead, the following kernel density estimator can be used:

$$(5.62) \qquad G_{ny}(y|z) = \frac{1}{nh_{ny}h_{nz} p_n(z)} \sum_{i=1}^{n} K_Y\left(\frac{Y_i - y}{h_{ny}}\right) K_Z\left(\frac{Z_{ni} - z}{h_{nz}}\right).$$

T_n is obtained by substituting (5.61) and (5.62) into (5.60).

Kernel estimators converge in probability at rates slower than $n^{-1/2}$ (see the Appendix). Therefore, $G_{ny}(y|z)/G_{nz}(y|z)$ is not $n^{1/2}$-consistent for $G_y(y|z)/G_z(y|z)$. However, integration over z and v in (5.59) creates an averaging effect that causes the integral and, therefore, T_n to converge at the rate $n^{-1/2}$. This is the reason for basing the estimator on (5.59) instead of (5.57). As was discussed in Chapter 2, a similar averaging effect takes place in density-weighted average derivative estimation of β in a single-index model and enables density-weighted average derivative estimators to converge at the rate $n^{-1/2}$. The formal statistical properties of T_n are discussed in Section 5.3.2

Now consider estimation of F. As in estimation of T, the derivation of the estimator takes place in two steps. The first step is to express F as a functional of T and the population distribution of (Y, X). The second step is to replace unknown population quantities in this expression with estimators. To take the first step, observe that because U is independent of X, $P(U \leq u | a < Z \leq b) = F(u)$ for any u and any a and b in the support of Z. Therefore, in particular, for any points y_1 and y_2 that are in the support of Y and satisfy $y_2 < y_1$,

$$F(u) = P[U \leq u | T(y_2) - u < Z \leq T(y_1) - u] .$$

$$(5.63) \qquad = A(u) / B(u) ,$$

where

$$(5.64) \qquad A(u) = E\{I(U \leq u) I[T(y_2) - u < Z \leq T(y_1) - u]\}$$

and

$$(5.65) \qquad B(u) = E\{I[T(y_2) - u < Z \leq T(y_1) - u]\} .$$

Equations (5.63)-(5.65) provide the desired expression for F in terms of T and the population distribution of (Y, X).

The estimator of F is obtained by replacing the unknown quantities in (5.63)-(5.65) with sample analogs. To do this, define $U_{ni} = T_n(Y_i) - Z_{ni}$, where $T_n(y)$ is replaced with an arbitrarily large negative number if $y < y_2$ and an arbitrarily large positive number if $y > y_1$. The estimator of $F(u)$ is

$$(5.66) \qquad F_n(u) = A_n(u) / B_n(u) ,$$

where

$$A_n(u) = \frac{1}{n} \sum_{i=1}^{n} I(U_{ni} \leq u) I[T_n(y_2) - u < Z_{ni} \leq T_n(y_1) - u]$$

and

$$B_n(u) = \frac{1}{n}\sum_{i=1}^{n} I[T_n(y_2) - u < Z_{ni} \leq T_n(y_1) - u].$$

The statistical properties of F_n are discussed in Section 5.3.2.

It may seem that F can be estimated more simply by the empirical distribution function of $U_n = T_n(Y) - Xb_n$, but this is not the case. T can be estimated $n^{1/2}$-consistently only over a compact interval $[y_2, y_1]$ that is a proper subset of the support of Y. This is because T may be unbounded at the boundaries of the support of Y (e.g, if the support of Y is $[0, \infty)$ and $T(y) = \log y$ and G_z is likely to be zero on the boundaries. Therefore, $T_n(Y)$ and U_n are $n^{1/2}$-consistently estimated only if $Y \in [y_2, y_1]$. In effect, F must be estimated from censored observations of U. The empirical distribution function of U_n is not a consistent estimator of F when there is censoring. Equation (5.66) provides an estimator that is consistent despite censoring.

5.3.2 Asymptotic Properties of T_n and F_n

This section gives conditions under which T_n is consistent for T, F_n is consistent for F, and $n^{1/2}(T_n - T)$ and $n^{1/2}(F_n - F)$ converge weakly to mean-zero Gaussian processes.

The following notation will be used. Let $\tilde{X}$ be a vector consisting of all components of X except the first. Define $q = \dim(X)$. Let $p(\bullet|\tilde{x})$ denote the probability density function of Z conditional on $\tilde{X} = \tilde{x}$. Let $r \geq 6$ and $s \geq 2$ be integers. Make the following assumptions:

Assumption 1: $\{Y_i, X_i: i = 1, ..., n\}$ is a random sample of (Y, X) in (5.1).

Assumption 2: (a) $|\beta_1| = 1$. (b) The distribution of the first component of X conditional on $\tilde{X} = \tilde{x}$ has a probability density for every $\tilde{x}$ in the support of $\tilde{X}$. (c) $\tilde{X}$ has bounded support.

Assumption 3: (a) U is independent of X and has a probability density. (b) Let f be the probability density function of U. There is an open subset I_U of the support of U such that $\sup\{f(u): u \in I_U\} < \infty$, $\inf\{f(u): u \in I_U\} > 0$, and the derivatives $d^k f(u)/du^k$ $(k = 1, ..., r + s)$ exist and are uniformly bounded over I_U.

Assumption 4: T is a strictly increasing, differentiable function everywhere on the support of Y.

Assumption 5: There are open intervals of the real line, I_Y and I_Z, such that (a) $y_0 \in I_Y$. (b) $y \in I_Y$ and $z \in I_Z$ implies $T(y) - z \in I_U$. (c) $p(z)$ and $p(z|\tilde{x})$ are bounded uniformly over $z \in I_Z$ and $\tilde{x}$ in the support of $\tilde{X}$. Moreover, $\inf\{p(z): z \in I_Z\} > 0$. (d) The derivatives $d^k p^k(z)/dz^k$ and $\partial^k p(z|\tilde{x})/\partial z^k$ (k = 1, ..., r + 1} exist and are uniformly bounded for all $z \in I_Z$ and $\tilde{x}$ in the support of $\tilde{X}$. (e) $T(y_0) = 0$, and $\sup\{|T(y)|: y \in I_Y\} < \infty$. For $k = 1, ..., r + 1$, the derivatives $dT^k(y)/dy^k$ exist and are uniformly bounded over $y \in I_Y$.

Assumption 6: S_w is compact, $S_w \subset I_Z$, (5.59) holds, and $d^k w(z)/dz^k$ (k = 1, ..., r + 1) exists and is bounded for all $z \subset I_z$.

Assumption 7: There is a $(q - 1)\times 1$ vector-valued function $\Omega(y, x)$ such that $E\Omega(Y,X) = 0$, the components of $E[\Omega(Y,X)\Omega(Y,X)']$ are finite, and as $n \to \infty$

$$n^{1/2}(b_n - \beta) = \frac{1}{n^{1/2}}\sum_{i=1}^{n} \Omega(Y_i, X_i) + o_p(1).$$

Assumption 8: K_Y has support [-1, 1], is bounded and symmetrical about 0, has bounded variation, and satisfies

$$\int_{-1}^{1} v^j K_Y(v)dv = \begin{cases} 1 & \text{if } j = 0 \\ 0 & \text{if } 1 \le j \le s-1 \\ \text{nonzero} & \text{if } j = s. \end{cases}$$

K_Z has support [-1, 1], is bounded and symmetrical about 0, and satisfies

$$\int_{-1}^{1} v^j K_Z(v)dv = \begin{cases} 1 & \text{if } j = 0 \\ 0 & \text{if } 1 \le j \le r-1 \\ \text{nonzero} & \text{if } j = r. \end{cases}$$

K_Z is everywhere twice differentiable. The derivatives are bounded and have bounded variation. The second derivative satisfies $|K_Z''(v_1) - K_Z''(v_2)| \le M|v_1 - v_2|$ for some $M < \infty$.

Assumption 9: As $n \to \infty$, $nh_{nz}^{2r} \to 0$, $nh_{ny}^{2s} \to 0$, $1/(nh_{nz}^{8}) \to 0$, and $(\log n)/(n^{1/2}h_{ny}^{1/2}h_{nz}^{3}) \to 0$.

Assumptions 1-6 specify the model and establish smoothness conditions implying, among other things, that G_y and G_z exist. Assumption 7 is satisfied by all of the estimators of β discussed in Chapter 2. Assumption 8 requires K_Z but not K_Y to be a higher-order kernel. A higher-order kernel is needed for K_Z because G_{nz} is a functional of derivatives of K_Z. Derivative functionals converge relatively slowly, and the higher-order kernel for K_Z is needed to insure sufficiently rapid convergence. Assumptions 8 and 9 are satisfied, for example, if K_Y is a second-order kernel, K_Z is a sixth-order kernel, $h_{ny} \propto n^{-1/3}$, and $h_{nz} \propto n^{-1/10}$.

The following theorems give the asymptotic properties of T_n and F_n:

Theorem 5.3: Let $[y_2, y_1] \in I_Y$. Under assumptions 1-9:

(a) $\operatorname*{plim}_{n\to\infty} \sup_{y_2 \le y \le y_1} |T_n(y) - T(y)| = 0$

(b) For $y \in [y_2, y_1]$, $n^{1/2}(T_n - T)$ converges weakly to a tight, mean-zero Gaussian process. ■

Theorem 5.4: Let $[u_0, u_1] \in I_U$ and $P[T(y_2) - u < Z \le T(y_1) - u] > 0$ whenever $u \in [u_0, u_1]$. Under assumptions 1-9:

(a) $\operatorname*{plim}_{n\to\infty} \sup_{u_0 \le u \le u_1} |F_n(u) - F(u)| = 0$

(b) For $u \in [u_0, u_1]$, $n^{1/2}(F_n - F)$ converges weakly to a tight, mean-zero Gaussian process. ■

Parts (a) of Theorems 5.3 and 5.4 establish uniform consistency of T_n and F_n. Parts (b) show, among other things, that the rates of convergence in probability of $T_n(y)$ and $F_n(u)$ are $n^{-1/2}$ and that the centered, normalized forms of T_n and F_n are asymptotically normally distributed. The expressions for the covariance functions of the limiting stochastic processes of $n^{1/2}(T_n - T)$ and $n^{1/2}(F_n - F)$ are very lengthy. They are given in Horowitz (1996b) together with methods for estimating the covariance functions consistently.

The proofs of Theorems 5.3 and 5.4 are also given in Horowitz (1996b). The proofs rely heavily on empirical process methods described by Pollard (1984). Roughly speaking, the proofs begin by using Taylor-series expansions to approximate $T_n - T$ by an integral of a linear functional of kernel estimators and to approximate $F_n - F$ by a linear functional of $A_n - A$ and $B_n - B$. In a second step, it is shown that these approximating functionals can, themselves, be approximated by empirical processes. Finally, uniform laws of large numbers and functional central limit theorems (central limit theorems for sums of random functions instead of sums of random variables) are used to establish the conclusions of the theorem.

5.4 Predicting *Y* Conditional on *X*

This section discusses how b_n, T_n and F_n can be used to make predictions of Y conditional on X.

The most familiar predictor of Y conditional on $X = x$ is a consistent estimator of $E(Y|X = x)$. Because (5.1) is a single-index model, an estimator of $E(Y|X = x)$ can always be obtained by carrying out the nonparametric mean-regression of Y on Xb_n. This estimator converges in probability to the true conditional expectation at a rate that is slower than $n^{-1/2}$. An estimator of the conditional expectation that converges in probability at the rate $n^{-1/2}$ can be obtained if T can be estimated with $n^{-1/2}$ accuracy over the entire support of Y. This is usually possible in the models of Section 5.1.1, where T is known up to a finite-dimensional parameter α. If a_n is a $n^{1/2}$-consistent estimator of α, and T_n^{-1} is the inverse of $T(y, a_n)$, and T is a differentiable function of each of its arguments, then $E(Y|X = x)$ is estimated $n^{1/2}$-consistently by

$$\frac{1}{n}\sum_{i=1}^{n} T_n^{-1}(U_{ni} + xb_n),$$

where b_n is a $n^{1/2}$-consistent estimator of β and $U_{ni} = T(Y_i\ a_n) - X_i b_n$.

Uniform $n^{1/2}$-consistent estimation of T usually is not possible in models where T is nonparametric, and $E(Y|X = x)$ cannot be estimated $n^{1/2}$-consistently in such models. A predictor that usually can be estimated $n^{1/2}$-consistently is the conditional median of Y or, possibly, some other conditional quantile. The γ quantile of Y conditional on $X = x$ $(0 < \gamma < 1)$ is

$$y_\gamma(x) = T^{-1}(x\beta + u_\gamma),$$

where u_γ is the γ-quantile of the distribution of U. To form a $n^{1/2}$-consistent estimator of $y_\gamma(x)$, let F_n be an estimator of F that is $n^{1/2}$-consistent in a neighborhood of u_γ. Let T_n be an estimator of T that is $n^{1/2}$-consistent in a neighborhood of $y_\gamma(x)$. In models where F is known, F can be used to estimate itself. Estimate u_γ by $u_{n\gamma} = \inf\{u\text{: } F_n(u) > \gamma\}$. Then $y_{n\gamma}(x)$ is estimated $n^{1/2}$-consistently by $y_{n\gamma}(x) = \inf\{y\text{: } T_n(y) > xb_n + u_{n\gamma}\}$. Cheng, Wei, and Ying (1997) and Horowitz (1996b) provide detailed discussions of the asymptotic distributional properties of $y_{n\gamma}(x)$ for the models discussed in Sections 5.2.5 and 5.3, respectively.

5.5 An Empirical Example

This section presents an empirical example that illustrates some of the techniques described in Sections 5.1-5.5. The data were assembled by Kennan (1985), who studied the relation between the durations of contract strikes and the level of economic activity. The data give the durations, in days, of 566 strikes involving 1000 or more workers in U.S. manufacturing during 1968-1976. The level of economic activity is measured by an index of industrial production in manufacturing (INDP). In the analysis reported here, high values of INDP indicate low levels of economic activity.

The analysis consists of using the data on strike durations and INDP to estimate three versions of model (5.1). In each version, Y is the duration of strike and X is INDP. The first estimated version of (5.1) is the semiparametric proportional hazards model of Section 5.2.1. In this model, T is nonparametric and $F(u) = 1 - \exp(-e^u)$. The second estimated version of (5.1) is a loglinear regression model. In this model, $T(y) = \log y$ and F is nonparametric. Estimation of this model is discussed in Section 5.1. In the third version of (5.1), T and F are both nonparametric. Estimation of this version of the model is discussed in Section 5.3. The version of (5.1) in which T and F are nonparametric nests both the proportional hazards model and the loglinear regression model. The proportional hazards and loglinear regression models are non-nested, however. There are versions of the proportional hazards model that are not loglinear regressions, specifically any proportional hazards model in which the integrated baseline hazard function does not have the form $\Lambda_0(y) = y^\alpha$ for some $\alpha > 0$. There are also loglinear regressions that are not proportional hazards models, specifically any loglinear model in which U does not have the extreme-value distribution.

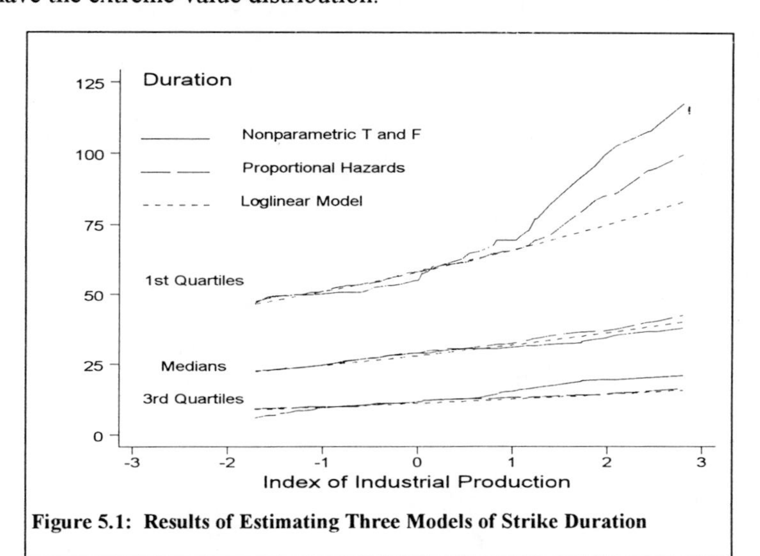

Figure 5.1: Results of Estimating Three Models of Strike Duration

The estimation results are summarized in Figure 5.1, which shows each model's estimates of the conditional first quartile, median, and third quartile of the distribution of strike durations given INDP. All of the models give results that are consistent with Kennan's (1985) finding that strike durations tend to be longer when at low levels of economic activity (high values of INDP) than at high levels. In addition, the conditional first-quartiles and medians are similar in all models. The conditional third quartiles are similar when INDP ≤ 1.25 but diverge at higher values of INDP. At high values of INDP, the estimated distribution of strike durations is more highly skewed to the right according to the model with nonparametric T and F than according to the proportional hazards model. The distribution of strike durations is less highly skewed to the right according to the loglinear model than according to the proportional hazards model.

Because the model with nonparametric T and F nests the other two models, these results suggest the possibility that the proportional hazards and loglinear models are misspecified, at least for large values of INDP. However, fewer than 7 percent of the data have values of INDP exceeding 1.25, and a formal test of the proportional hazards and loglinear models against the more general model is not yet available. Therefore, one cannot rule out the possibility that the differences among the estimated third quartiles are artifacts of random sampling error. Nonetheless, the estimation results demonstrate that even a relatively complicated semiparametric estimator such as that of Section 5.3 can yield useful estimates with samples of the sizes found in applications.

Appendix: Nonparametric Estimation

This appendix summarizes properties of kernel nonparametric density and mean-regression estimators that are used in the text. See Härdle (1990) and Silverman (1986) for more detailed presentations.

A.1 Nonparametric Density Estimation

Let X be a random variable with unknown probability density function $p(\bullet)$. Let $\{X_i: i = 1, \ldots, n\}$ be a random sample of X. The problem addressed in this section is to estimate p consistently from $\{X_i\}$ without assuming that p belongs to a finite-dimensional parametric family of functions. Assume for now that X is a scalar. The case of multidimensional X is treated in Section A.1.1.

Let P denote the cumulative distribution function (CDF) of X. For any real x, $P(x)$ is estimated consistently by the empirical distribution function of X, which is

$$P_n(x) = \frac{1}{n}\sum_{i=1}^{n} I(X_i \le x).$$

$P(x)$ and $p(x)$ are related by $p(x) = dP(x)/dx$, but $p(x)$ cannot be estimated by $dP_n(x)/dx$ because P_n is a step function. This problem can be solved by adding to each observation X_i a random number $h_n\varepsilon_i$, where ε_i is sampled randomly from a distribution with known density K, and h_n is a scale parameter. The distribution of the resulting random variable, $X_i + h_n\varepsilon_i$, is the convolution of the empirical distribution of X and the distribution of $h_n\varepsilon$. The convolution distribution is smooth and has the density function

$$p_n(x) = \frac{1}{h_n}\int K\left(\frac{x-z}{h_n}\right) dP_n(z)$$

$$= \frac{1}{nh_n} \sum_{i=1}^{n} K\left(\frac{x - X_i}{h_n}\right).$$

The function p_n is called a *kernel nonparametric density estimator*. K is called the *kernel* function. Intuition suggests that $p_n(x)$ should be close to $p(x)$ if n is large (so that P_n is close to P) and h_n is small (so that the convolution with $h_n\varepsilon$ does not alter the empirical distribution of X too much). Conditions under which this intuition is correct will now be given.

Suppose that K is bounded and satisfies

$$\int_{-\infty}^{\infty} K(v)dv = 1,$$

$$\int_{-\infty}^{\infty} vK(v)dv = 0,$$

$$\int_{-\infty}^{\infty} v^2 K(v)dv = A < \infty,.$$

and

$$\int [K(v)]^2 dv = B.$$

These conditions are satisfied if, for example, K is a bounded probability density function. Suppose, also, that $h_n \to 0$ and $nh_n/(\log n) \to \infty$ as $n \to \infty$. Then it can be proved that if p is uniformly continuous,

$$\text{(A.1)} \qquad \lim_{n \to \infty} \sup_x |p_n(x) - p(x)| = 0$$

almost surely. See Silverman (1978). Thus, p_n is a strongly uniformly consistent estimator of p.

Now suppose that p is twice continuously differentiable. Then for each x, the fastest possible rate of convergence in probability of $p_n(x)$ to $p(x)$ is $n^{-2/5}$ (Stone 1980). This occurs when $h_n = cn^{-1/5}$ for some $c > 0$. Moreover,

$$\text{(A.2)} \qquad n^{2/5}[p_n(x) - p(x)] \to^d N(\mu, \sigma^2),$$

where

$$\mu = \frac{1}{2} c^2 A p''(x)$$

and

$$\sigma^2 = \frac{B}{c} p(x) .$$

The asymptotic mean-square error (AMSE) of $p_n(x)$ is $\mu + \sigma^2$. Assuming that $p''(x) \neq 0$, the AMSE is minimized by setting

$$c = \left[\frac{Bf(x)}{A^2 p''(x)^2} \right]^{1/5}$$

The integrated mean-square error of p_n is

$$IMSE = E\int [p_n(x) - p(x)]^2 dx .$$

This is minimized asymptotically by setting

$$\text{(A.3)} \qquad c = \left[\frac{B}{A^2 \int p''(v)^2 dv} \right]^{1/5} .$$

The uniform rate of convergence of p_n to p is slower than the pointwise rate of $n^{-2/5}$. It can be proved that if p is uniformly continuous, then

$$\left(\frac{nh_n}{\log n} \right)^{1/2} \sup_x |p_n(x) - p(x)| = O(1)$$

almost surely (Silverman 1978).

Equation (A.2) shows that the asymptotic distribution of $n^{2/5}[p_n(x) - p(x)]$ is not centered at 0. Thus $p_n(x)$ is asymptotically biased. The asymptotic bias can be removed at the cost of a slower rate of convergence by setting $h_n \propto n^{-r}$ where $r > 1/5$. This is called *undersmoothing*. With undersmoothing, the pointwise rate of convergence of $p_n(x)$ is $n^{-(1-r)/2}$, and

$$(nh_n)^{1/2} [p_n(x) - p(x)] \to^d N[0, Bp(x)] .$$

Asymptotic bias can be removed while keeping the rate of convergence $n^{-2/5}$ by using a jackknife-like method proposed by Schucany and Sommers (1977). To obtain this estimator, let $p_n(x, h_{n1})$ be the kernel density estimator of $p(x)$ based on the bandwidth $h_{n1} = cn^{-1/5}$. Let $p_n(x, h_{n2})$ be the estimator based on bandwidth $h_{n2} = cn^{-\delta/5}$, where $0 < \delta < 1$. Define

$$\hat{p}_n(x) = \frac{p_n(x, h_{n1}) - (h_{n1}/h_{n2})^2 p_n(x, h_{n2})}{1 + \dfrac{h_{n1}}{h_{n2}}}.$$

Then

$$(nh_{n1})^{1/2}[\hat{p}_n(x) - p(x)] \to^d N[0, Bf(x)].$$

The rate of convergence of p_n can be increased if p has more than two continuous derivatives through the use of a *higher-order kernel*. A kernel K of order s satisfies

$$\int_{-\infty}^{\infty} K(v)dv = 1,$$

$$\int_{-\infty}^{\infty} v^j K(v)dv = 0, \quad j = 1, 2, ..., s-1$$

$$\int_{-\infty}^{\infty} v^s K(v)dv = A < \infty,$$

and

$$\int [K(v)]^2 dv = B.$$

If K is a probability density function, then $s = 2$. A higher-order kernel has $s > 2$. A higher-order kernel must be negative at some values of its argument in order to have a second "moment" of zero. Therefore, a probability density function cannot be a higher-order kernel, but higher-order kernels can be constructed by using formulae provided by Müller (1984). The following is a 4'th order kernel with support [-1, 1]:

$$K(v) = \frac{105}{64}(1 - 5v^2 + 7v^4 - 3v^6)I(|v| \le 1).$$

Suppose that p has s continuous derivatives, $h_n = cn^{-1/(2s+1)}$ for some $c > 0$, and p_n is the estimator of that is obtained by using a kernel K of order s and bandwidth h_n. Then the rate of convergence of p_n to p is $n^{2s/(2s+1)}$. This rate increases as s increases, so a faster rate of convergence can be obtained by using a higher-order kernel if p has the required derivatives. If p is s times continuously differentiable, K is an order s kernel, and $h_n = cn^{-1/(2s+1)}$, then

$$n^{s/(2s+1)}[p_n(x) - p(x)] \to^d N(\mu, \sigma^2),$$

where

$$\mu = \frac{1}{s!}c^s A p^{(s)}(x)$$

and

$$\sigma^2 = \frac{B}{c}p(x).$$

The AMSE of $p_n(x)$ is minimized by choosing c to minimize $\mu^2 + \sigma^2$. Assuming that $p^{(s)}(x) \neq 0$, the result is

$$c = \left\{ \frac{Bp(x)}{2s} \left[\frac{s!}{Ap^{(s)}(x)} \right]^2 \right\}^{1/(2s+1)}.$$

The asymptotic integrated mean-square error is minimized by setting

$$c = \left\{ \frac{B}{2s} \left(\frac{s!}{A} \right)^2 \frac{1}{\int \left[p^{(s)}(v) \right]^2 dv} \right\}^{1/(2s+1)}.$$

In applications, the bandwidth that minimizes the asymptotic integrated mean-square error, $h_{n,opt}$, can be estimated by the method of least-squares cross validation. To describe this method, let $p_{nh}(x)$ be the estimate of $p(x)$ that is obtained by using bandwidth h, and let $p_{nhi}(x)$ be the estimate that is obtained by using bandwidth h and omitting observation X_i from the data. Define

$$T(h) = \int_{-\infty}^{\infty} [p_{nh}(x)]^2\, dx - \frac{2}{n} \sum_{i=1}^{n} p_{nhi}(X_i)\,.$$

The least-squares cross-validation estimator of $h_{n,opt}$ is the value of h that minimizes $T(h)$.

The least-squares cross-validation estimator, $h_{n,LCV}$, is easy to compute, but it converges to $h_{n,opt}$ at the slow rate of $n^{-1/10}$. That is, $h_{n,LCV}/h_{n,opt} = O_p(n^{-1/10})$. Consequently, $h_{n,LCV}$ can be noisy; in samples of moderate size; $h_{n,LCV}$ may differ greatly from $h_{n,opt}$. Jones, *et al.* (1996) describe other bandwidth estimators that are more difficult to compute but have faster rates of convergence and, therefore, are likely to be less noisy in samples of moderate size.

A.1.1 Density Estimation when X Is Multidimensional

In this section, it is assumed that X is a d-dimensional random variable with $d > 1$. Let p be the probability density function of X, K be a kernel function of a d-dimensional argument. Such a function could be a multivariate probability density function or the product of univariate kernel functions. Let $\{h_n\}$ be a sequence of bandwidths. The kernel nonparametric estimator of $p(x)$ is

$$p_n(x) = \frac{1}{nh_n^d} \sum_{i=1}^{n} K\left(\frac{x - X_i}{h_n}\right).$$

It is possible to use different bandwidths for different components of X, but this refinement will not be pursued here.

If $h_n \to 0$ and $nh_n^d/(\log n) \to \infty$ and $n \to \infty$, then (A.1) holds when X is multidimensional. To obtain the rate of convergence and asymptotic distribution of $p_n(x) - p(x)$, it is necessary to define the multivariate generalization of a kernel of order s. To this end, let $j = (j_1, \ldots, j_d)$ be a d-dimensional vector whose components are all non-negative integers. Let v be the d-dimensional vector whose components are $(v_1, \ldots, v_d)$. Define

$$v^j = \prod_{k=1}^{d} v_k^{j_k}$$

and

$$|j| = \sum_{k=1}^{d} j_k\,.$$

Let $K(v) = K(v_1, \ldots, v_d)$ and $dv = dv_1 dv_2 \ldots dv_d$. Then K is an order s kernel if

$$\int_{-\infty}^{\infty} K(v)dv = 1,$$

$$\int_{-\infty}^{\infty} v^j K(v)dv = 0, \quad |j| = 1, 2, \ldots, s-1$$

$$\int_{-\infty}^{\infty} v_k^s K(v)dv = A < \infty, \text{ for each } k = 1, \ldots, d$$

and

$$\int [K(v)]^2 dv = B.$$

A d-dimensional kernel K of order s can be obtained by setting

$$K(v) = \prod_{k=1}^{d} K_1(v_k),$$

where K_1 is an order-s kernel of a one-dimensional argument.

Now assume that the mixed partial derivatives of p of order up to s exist and are continuous. Let K be an order s kernel. Then the fastest possible rate of convergence in probability of $p_n(x)$ to $p(x)$ is $n^{-s/(2s+d)}$. This rate decreases as d increases. As a result, the sample size needed to achieve a given estimation precision increases very rapidly as d increases. This is the *curse of dimensionality* (Huber 1985). Silverman (1986) gives an illustrative example in which it is necessary to increase n by a factor of nearly 200 to keep the estimation precision constant when d increases from 1 to 5.

When p is s times continuously differentiable and K is an order s kernel, the fastest possible rate of convergence is achieved when $h_n = cn^{-1/(2s+d)}$ for some c > 0 (Stone 1980). Under these conditions,

$$n^{s/(2s+d)}[p_n(x) - p(x)] \to^d N(\mu, \sigma^2),$$

where

$$\mu = \frac{c^s}{s!} A \sum_{k=1}^{d} \frac{\partial^s p(x)}{\partial x_k^s},$$

x_k is the k'th component of x, and

$$\sigma^2 = \frac{B}{c^d} p(x).$$

The asymptotic mean-square error of $p_n(x)$ is minimized by setting

$$c = \left\{ \frac{dBp(x)(s!)^2}{2s} \left[A \sum_{k=1}^{d} \frac{\partial^s p(x)}{\partial x_k^s} \right]^{-2} \right\}^{1/(2s+d)}.$$

The asymptotic integrated mean-square error of p_n is minimized by setting

$$c = \left\{ \frac{dB(s!)^2}{2sA^2 \int \left[\sum_{k=1}^{d} \frac{\partial^s p(x)}{\partial x_k^s} \right]^2 dv} \right\}^{1/(2s+d)}.$$

As in the one-dimensional case, the bandwidth that minimizes the asymptotic integrated mean-square error for $d > 1$ can be estimated by least-squares cross validation, among other ways.

A.1.2 Estimating Derivatives of a Density

Return to the case in which X is a scalar. Let $p^{(k)}(x)$ denote the k'th derivative of p at the point x. An obvious estimator of $p^{(k)}(x)$ is the k'th derivative of $p_n(x)$ or $p_n^{(k)}(x)$. This section summarizes some important properties of $p_n^{(k)}(x)$.

Differentiation of p_n shows that

$$p_n^{(k)}(x) = \frac{1}{nh_n^{k+1}} \sum_{i=1}^{n} K^{(k)}\left(\frac{x - X_i}{h_n} \right),$$

where $K^{(k)}$ is the k'th derivative of K and is assumed to exist everywhere. If K is an order s kernel that is k times continuously differentiable everywhere and if p is $k + s$ times continuously differentiable, then

$$E[p_n^{(k)}(x)] = p^{(k)}(x) + \frac{h_n^s}{s!} A p^{(k+s)}(x) + o(h_n^s),$$

and

$$\mathrm{Var}\left[p_n^{(k)}(x)\right] = \frac{B_k}{nh_n^{2k+1}} p(x),$$

where

$$B_k = \int K^{(k)}(v)^2 \, dv.$$

It follows from Chebyshev's inequality that $p_n^{(k)}(x)$ converges in probability to $p^{(k)}(x)$ if $h_n \to 0$ and $nh_n^{2k+1} \to \infty$ as $n \to \infty$.

It can be seen that for any given bandwidth sequence and order of kernel, the bias of $p_n^{(k)}(x)$ converges to 0 at the same rate as the bias of $p_n(x)$, but the variance of $p_n^{(k)}(x)$ converges more slowly than the variance of $p_n(x)$. The fastest possible rate of convergence of $p_n^{(k)}(x)$ is achieved when the square of the bias and the variance of $p_n^{(k)}(x)$ converge to zero at the same rate. This occurs when $h_n \propto n^{-1/(2s+2k+1)}$. The resulting rate of convergence is $n^{-s/(2s+2k+1)}$. Thus, estimators of derivatives of densities converge more slowly than estimators of densities themselves and require slower converging bandwidth sequences.

A.2 Nonparametric Mean Regression

Now consider the problem of estimating the conditional mean function $g(x) \equiv E(Y|X = x)$. Assume that X is a scalar, continuously distributed random variable. Let $\{Y_i, X_i: i = 1, ..., n\}$ be a random sample of (Y, X). Let $p_n(x)$ be the kernel nonparametric estimator of $p(x)$, the probability density of X at x, based on the sample, kernel K and bandwidth h_n. The kernel nonparametric estimator of $g(x)$ is

$$g_n(x) = \frac{1}{nh_n p_n(x)} \sum_{i=1}^{n} Y_i K\left(\frac{x - X_i}{h_n}\right).$$

Define $\sigma^2(x) = \mathrm{Var}(Y|x = x)$. Assume that $\sigma^2(x) < \infty$ for all finite x and that $E[\sigma^2(X)] < \infty$. If p and g are both s times continuously differentiable and K is an order s kernel, then the fastest possible rate of convergence in probability of g_n to g is $n^{-2/(2s+1)}$. As in nonparametric density estimation, the rate of convergence of a nonparametric mean-regression estimator can be increased if the necessary derivatives of p and g exist by using a higher-kernel.

The fastest possible rate of convergence in probability of g_n to g occurs when $h_n = cn^{-1/(2s+1)}$ for some $c > 0$ (Stone 1980, 1982). With this bandwidth, an order s kernel, and assuming existence of the relevant derivatives of p and g,

$$n^{s/(2s+1)}[g_n(x) - g(x)] \to^d N(\mu_R, \sigma_R^2),$$

where

$$\mu_R = \frac{c^s}{p(x)} AD(x),$$

$$D(x) = \sum_{k=1}^{s} \frac{1}{k!} \frac{d^k}{dv^k} \{[g(v+x) - g(x)]p(x)\}_{v=0},$$

and

$$\sigma_R^2 = \frac{B\sigma^2(x)}{cp(x)}.$$

The asymptotic bias of $g_n(x)$ can be removed without reducing the rate of convergence in probability by using a method analogous to that of Schucany and Sommers (1978) for nonparametric density estimation. See Härdle (1986) and Bierens (1987) for details.

As in nonparametric density estimation, c can be chosen to minimize either the asymptotic mean-square error of $g_n(x)$ or the (asymptotic) integrated mean-square error. The integrated mean-square error is

$$IMSE = E\int w(x)[g_n(x) - g(x)]^2 dx,$$

where w is any non-negative function satisfying

$$\int w(x)dx = 1$$

and for which the integral in *IMSE* exists. The asymptotic mean-square error of $g_n(x)$ is minimized by setting

$$c = \left[\frac{Bp(x)\sigma^2(x)}{2sA^2 D(x)^2}\right]^{1/(2s+1)}.$$

The asymptotic integrated mean-square error is minimized by setting

$$c = \left\{ \frac{1}{2s} \frac{A^2 \int w(v)[D(v)/p(v)]^2 dv}{B \int w(v)[\sigma^2(v)/p(v)]dv} \right\}^{1/(2s+1)}.$$

The bandwidth that minimizes the asymptotic integrated mean-square error can be estimated by minimizing the following cross-validation criterion function:

$$T_R(h) = \frac{1}{n} \sum_{i=1}^{n} w(X_i)[Y_i - g_{nhi}(X_i)]^2,$$

where $g_{nhi}(x)$ is the kernel estimator of $g(x)$ that is obtained by using bandwidth h and by omitting the observation (Y_i, X_i) from the sample.

The uniform rate of convergence of g_n to g is $[(\log n)/(nh_n)]^{1/2}$ almost surely. Specifically,

$$\left(\frac{nh_n}{\log n} \right)^{1/2} \sup_{x \in S} |g_n(x) - g(x)| = O(1)$$

almost surely, where S is any compact subset of the support of X on which p and g are s times continuously differentiable.

If X is d-dimensional with $d > 1$, then $g(x)$ can be estimated by

$$g_n(x) = \frac{1}{nh_n{}^d p_n(x)} \sum_{i=1}^{n} Y_i K\left(\frac{x - X_i}{h_n} \right),$$

where K is a kernel function of a d-dimensional argument. As nonparametric density estimation, the fastest possible rate of convergence in probability of $g_n(x)$ to $g(x)$ is $n^{-s/(2s+d)}$ when p and g are s times continuously differentiable. Thus, the curse of dimensionality arises in nonparametric mean-regression as in nonparametric density estimation. The fastest possible rate of convergence is achieved when $h_n = cn^{-1/(2s+d)}$ for some $c > 0$, in which case $n^{s/(2s+d)}[g_n(x) - g(x)]$ is asymptotically normally distributed.

Derivatives $g(x)$ can be estimated by differentiating $g_n(x)$. As in the case of density estimation, estimators of derivatives of g converge more slowly than the estimator of g itself, and slower converging bandwidth sequences are needed.

References

Ai, C. (1997). A Semiparametric Maximum Likelihood Estimator, *Econometrica*, **65**, 933-963.

Ai, C. and D. McFadden (1997). Estimation of Some Partially Specified Nonlinear Models, *Journal of Econometrics*, **76**, 1-37.

Amemiya, T. (1985). *Advanced Econometrics*, Cambridge, MA: Harvard University Press.

Amemiya, T. and J. L. Powell (1981). A Comparison of the Box-Cox Maximum Likelihood Estimator and the Non-Linear Two-Stage Least Squares Estimator, *Journal of Econometrics*, **17**, 351-381.

Andersen, P. K. and R. D. Gill (1982). Cox's Regression Model for Counting Processes: A Large Sample Study, *Annals of Statistics*, **10**, 1100-1120.

Andrews, D. W. K. and Y. J. Whang (1990). Additive Interactive Regression Models: Circumvention of the Curse of Dimensionality, *Econometric Theory*, **6**, 466-479.

Bennett, S. (1983a). Analysis of Survival Data by the Proportional Odds Model, *Statistics in Medicine*, **2**, 273-277.

Bennett, S. (1983b). Log-Logistic Regression Models for Survival Data, *Applied Statistics*, **32**, 165-171.

Beran, R. (1988). Prepivoting Test Statistics: A Bootstrap View of Asymptotic Refinements, *Journal of the American Statistical Association*, **83**, 687-697.

Bickel, P. J. and K. A. Doksum (1981). An Analysis of Transformations Revisited, *Journal of the American Statistical Association*, **76**, 296-311.

Bickel, P. J., C. A. J. Klaassen, Y. Ritov, and J. A. Wellner (1993). *Efficient and Adaptive Estimation for Semiparametric Models*, Baltimore: The Johns Hopkins University Press.

Bierens, H. J. (1987). Kernel Estimators of Regression Functions, in *Advances in Econometrics: 5th World Congress*, vol. 1, T. F. Bewley, ed., Cambridge: Cambridge University Press.

Bierens, H. J. and J. Hartog (1988). Non-Linear Regression with Discrete Explanatory Variables, with an Application to the Earnings Function, *Journal of Econometrics*, **38**, 269-299.

Billingsley, P. (1968). *Convergence of Probability Measures*, New York: John Wiley & Sons.

Box, G. E. P. and D. R. Cox (1964). An Analysis of Transformations, *Journal of the Royal Statistical Society*, Series B, **26**, 211-252.

Breiman, L. and J. H. Friedman (1985). Estimating Optimal Transformations for Multiple Regression and Correlation, *Journal of the American Statistical Association*, **80**, 580-598.

Carroll, R. J. (1982). Adapting for Heteroscedasticity in Linear Models, *Annals of Statistics*, **10**, 1224-1233.

Cavanagh, C. L (1987). Limiting Behavior of Estimators Defined by Optimization, unpublished manuscript, Department of Economics, Harvard University, Cambridge, MA.

Chamberlain, G. (1986). Asymptotic Efficiency in Semiparametric Models with Censoring, *Journal of Econometrics*, **32**, 189-218.

Charlier, E. (1994). A Smoothed Maximum Score Estimator for the Binary Choice Panel Data Model with Individual Fixed Effects and Application to Labour Force Participation, discussion paper no. 9481, CentER for Economic Research, Tilburg University, Tilburg, The Netherlands.

Charlier, E., B. Melenberg, and A. van Soest (1995). A Smoothed Maximum Score Estimator for the Binary Choice Panel Data Model with an Application to Labour Force Participation, *Statistica Neerlandica*, **49**, 324-342.

Cheng, S. C., L. J. Wei, and Z. Ying (1995). Analysis of Transformation Models with Censored Data, *Biometrika*, **82**, 835-845.

Cheng, S. C., L. J. Wei, and Z. Ying (1997). Predicting Survival Probabilities with Semiparametric Transformation Models, *Journal of the American Statistical Association*, **92**, 227-235.

Clayton, D. and Cuzick, J. (1985). Multivariate Generalizations of the Proportional Hazards Model, *Journal of the Royal Statistical Society*, Series A, **148**, 82-117.

Cosslett, S. R. (1981). Efficient Estimation of Discrete-Choice Models, in *Structural Analysis of Discrete Data with Econometric Applications*, D. F. Manski and D. McFadden, eds., Cambridge, MA: MIT Press.

Cosslett, S. R. (1987). Efficiency Bounds for Distribution-Free Estimators of the Binary Choice and Censored Regression Models, *Econometrica*, **55** 559-586.

Cox, D. R. (1972). Regression Models and Life tables, *Journal of the Royal Statistical Society*, Series B, **34**, 187-220.

Davidson, R. and J. G. MacKinnon (1993). *Estimation and Inference in Econometrics*, New York: Oxford University Press.

Dempster, A. P., N. M. Laird, and D. R. Rubin (1977). Maximum Likelihood Estimation from Incomplete Data Via the EM Algorithm, *Journal of the Royal Statistical Society*, Series B, **39**, 1-38.

Elbers, C. and G. Ridder (1982). True and Spurious Duration Dependence: The Identifiability of the Proportional Hazard Model, *Review of Economic Studies*, **49**, 403-409.

Fischer, G. W. and D. Nagin (1981). Random Versus Fixed Coefficient Quantal Choice Models, in *Structural Analysis of Discrete Data with Econometric Applications*, D. F. Manski and D. McFadden, eds., Cambridge, MA: MIT Press.

Gallant, A. R. 91987). *Nonlinear Statistical Models*, New York: John Wiley & Sons.

Gørgens, T. and J. L. Horowitz (1996). Semiparametric Estimation of a Censored Regression Model with an Unknown Transformation of the Dependent Variable, working paper no. 95-15, Department of Economics, University of Iowa, Iowa City, IA.

Gronau, R. (1994). Wage Comparisons - A Selectivity Bias," *Journal of Political Economy*, **82**, 1119-1143.

Hall, P. (1986). On the Bootstrap and Confidence Intervals, *Annals of Statistsics*, **14**, 1431-1452.

Hall, P. (1992). *The Bootstrap and Edgeworth Expansion*, New York: Springer-Verlag.

Hall, P. and Ichimura, H. (1991). Optimal Semi-Parametric Estimation in Single-Index Models, working paper no. CMA-SR5-91, Centre for Mathematics and Its Applications, Australian National University, Canberra, Australia.

Hansen, L. P. (1982). Large Sample Progerties of Generalized Method of Moments Estimators, *Econometrica*, **50**, 1029-1054.

Härdle, W. (1986). A Note on Jackknifing Kernel Regression Function Estimators, *IEEE Transactions of Information Theory*, **32**, 298-300.

Härdle, W. (1990). *Applied Nonparametric Regression*, Cambridge: Cambridge University Press

Härdle, W. and O. Linton (1994). Applied Nonparametric Methods, in *Handbook of Econometrics*, vol. 4, R. F. Engle and D. F. McFadden, eds., Amsterdam: Elsevier, Ch. 38.

Härdle, W. and T. M. Stoker (1989). Investigating Smooth Multiple Regression by the Method of Average Derivatives, *Journal of the American Statistical Association*, **84**, 986-995.

Härdle, W. and A. B. Tsybakov (1993). How Sensitive Are Average Derivatives? *Journal of Econometrics*, **58**, 31-48.

Hastie, T.J. and R. J. Tibshirani (1990). *Generalized Additive Models*, London: Chapman and Hall.

Hausman, J. A. and D. A. Wise (1978). A Conditional Probit Model for Qualitative Choice: Discrete Decisions Recognizing Interdependence and Heterogeneous Preferences, *Econometrica*, **46**, 403-426.

Heckman, J. J. (1974). Shadow Prices, Market Wages, and Labor Supply, *Econometrica*, **42**, 679-693.

Heckman, J. J. (1981a). Statistical Models for Discrete Panel Data, in *Structural Analysis of Discrete Data with Econometric Applications*, D. F. Manski and D. McFadden, eds., Cambridge, MA: MIT Press.

Heckman, J. J. (1981b). The Incidental Parameters Problem and the Problem of Initial Conditions in Estimating a Discrete Time-Discrete Data Stochastic Process, in *Structural Analysis of Discrete Data with Econometric Applications*, D. F. Manski and D. McFadden, eds., Cambridge, MA: MIT Press.

Heckman, J. and B. Singer (1984a). The Identifiability of the Proportional Hazard Model, *Review of Economics Studies*, **51**, 231-243.

Heckman, J. and B. Singer (1984b). A Method for Minimizaing the Impact of Distributional Assumptions in Econometric Models for Duration Data, *Econometrica*, **52**, 271-320,.

Honoré, B. E. (1990). Simple Estimation of a Duration Model with Unobserved Heterogeneity, *Econometrica*, **58**, 453-473.

Horowitz, J. L. (1992). A Smoothed Maximum Score Estimator for the Binary Response Model, *Econometrica*, **60**, 505-531.

Horowitz, J. L. (1993a). Semiparametric Estimation of a Work-Trip Mode Choice Model, *Journal of Econometrics*, **58**, 49-70.

Horowitz, J. L. (1993b). Semiparametric and Nonparametric Estimation of Quantal Response Models, in *Handbook of Statistics, Vol 11*, G. S. Maddala, C. R. Rao, and H. D. Vinod, eds., Amsterdam: North-Holland Publishing Company.

Horowitz, J. L. (1993c). Optimal Rates of Convergence of Parameter Estimators in the Binary Response Model with Weak Distributional Assumptions, *Econometric Theory*, **9**, 1-18.

Horowitz, J. L. (1996a). Bootstrap Critical Values for Tests Based on the Smoothed Maximum Score Estimator, working paper no. 96-02, Department of Economics, University of Iowa, Iowa City, IA.

Horowitz (1996b). Semiparametric Estimation of a Regression Model with an Unknown Transformation of the Dependent Variable, *Econometrica*, **64**, 103-137.

Horowitz, J. L. (1997). Bootstrap Methods in Econometrics: Theory and Numerical Performance, in *Advances in Economics and Econometrics: Theory and Applications*, Vol. III, D. M. Kreps and K. F. Wallis, eds., Cambridge: Cambridge University Press, Ch. 7.

Horowitz, J. L. and Härdle, W. (1996). Direct Semiparametric Estimation of Single-Index Models with Discrete Covariates, *Journal of the American Statistical Association*, **91**, 1632-1640.

Hougaard, P. (1984). Life Table Methods for Heterogeneous Populations: Distributions Describing the Heterogeneity, *Biometrika*, **61**, 75-83.

Hougaard, P. (1986). Survival Models for Heterogeneous Populations Derived from Stable Distributions, *Biometrika*, **73**, 387-396.

Hsieh, D., C. F. Manski, and D. McFadden (1985). Estimation of Response Probabilities from Augmented Retrospective Observations, *Journal of the American Statistical Association*, **80**, 651-662.

Huber, P. J. (1985). Projection Pursuit, *Annals of Statistics*, **13**, 435—475.

Ibragimov, I. A. and R. Z. Has'minskii (1981). *Statistical Estimation: Asymptotic Theory*, New York: Springer-Verlag.

Ichimura, H. (1993). Semiparametric Least Squares (SLS) and Weighted SLS Estimation of Single-Index Models, *Journal of Econometrics*, **58**, 71-120.

Ichimura, H. and L.-F. Lee (1991). Semiparametric Least Squares Estimation of Multiple Index Models: Single Equation Estimation, in *Nonparametric and Semiparametric Methods in Econometrics and Statistics*, W. A. Barnett, J. Powell, and G. Tauchen, eds., Cambridge: Cambridge University Press, Ch. 1.

Imbens, G. (1992). An Efficient Method of Moments Estimator for Discrete Choice Models with Choice-Based Sampling, *Econometrica*, **60**, 1187-1214.

Jennrich, R. I. (1969). Asymptotic Properties of Non-Linear Least Squares Estimators, *Annals of Mathematical Statistics*, **40**, 633-643.

Jones, M. C., J. S. Marron, and S. J. Sheather (1996). A Brief Survey of Bandwidth Selection for Density Estimation, *Journal of the American Statistical Association*, 91, 401-407.

Kennan, J. (1985). The Duration of Contract Strikes in U.S. Manufacturing, *Journal of Econometrics*, **28**, 5-28.

Kiefer, J. and J. Wolfowitz (1956). Consistency of the Maximum Likelihood Estimator in the Presence of Infinitely Many Incidental Parameters, *Annals of Mathematical Statistics*,**27**, 887-906.

Kim, J. and D. Pollard (1990). Cube Root Asymptotics, *Annals of Statistics*, **18**, 191-219.

Klein, R. W. and R. H. Spady (1993). An Efficient Semiparametric Estimator for Binary Response Models, *Econometrica*, **61**, 387-421.

Kooreman, P.. and B. Melenberg (1989). Maximum Score Estimation in the Ordered Response Model, discussion paper no. 8948, CentER for Economic Research, Tilburg University, Tilburg, The Netherlands.

Lam, K. F. and A. Y. C. Kuk (1997). A Marginal Likelihood Approach to Estimation in Frailty Models, *Journal of the American Statistical Association*, **92**, 985-990.

Lancaster, T. (1979). Econometric Methods for the Duration of Unemployment, *Econometrica*, **47**, 939-956.

Lee, M.-J (1992). Median Regression for Ordered Discrete Response, *Journal of Econometrics*, **51**, 59-77.

Li, K.-C. (1991). Sliced Inverse Regression for Dimension Reduction, *Journal of the American Statistical Association*, **86**, 316-342.

Linton, O. B. (1996). Efficient Estimation of Additive Nonparametric Regression Models, *Biometrika*, forthcoming.

Linton, O. B. and W. Härdle (1996). Estimating Additive Regression Models with Known Links, *Biometrika*, **83**, 529-540.

Linton, O. B. and J. B. Nielsen (1995). A Kernel Method of Estimating Structured Nonparametric Regression Based on Marginal Integration, *Biometrika*, **82**, 93-100.

MacKinnon, J. G. and L. Magee (1990). Transforming the Dependent Variable in Regression Models, *International Economic Review*, **31**, 315-339.

Manski, C. F. (1985). Semiparametric Analysis of Discrete Response: Asymptotic Properties of the Maximum Score Estimator, *Journal of Econometrics*,**27**, 313-334.

Manski, C. F. (1987). Semiparametric Analysis of Random Effects Linear Models from Binary Panel Data, *Econometrica*, **55**, 357-362.

Manski, C. F. (1988). Identification of Binary Response Models, *Journal of the American Statistical Association*, **83**, 729-738.

Manski, C.F. (1994). The Selection Problem, in *Advances in Econometrics: Sixth World Congress*, vol. 1, C. A. Sims, ed., Cambridge: Cambridge University Press, Ch. 4.

Manski, C. F. (1995). *Identification Problems in the Social Sciences*, Cambridge, MA: Harvard University Press.

Manski, C. F. and S. Lerman (1977). The Estimation of Choice Probabilities from Choice-Based Samples, *Econometrica*, **45**, 1977-1988.

Manski, C. F. and D. McFadden (1981). Alternative Estimators and Sample Designs for Discrete Choice Analysis, in *Structural Analysis of Discrete Data with Econometric Applications*, D. F. Manski and D. McFadden, eds., Cambridge, MA: MIT Press.

Manski, C. F. and T. S. Thomson (1986). Operational Characteristics of Maximum Score Estimation, *Journal of Econometrics*, **32,** 65-108.

Matzkin, R. L. (1992). Nonparametric and Distribution-Free Estimation of the Binary Threshold Crossing and the Binary Choice Models, *Econometrica*, **60**, 239-270.

Matzkin, R.L. (1994). Restrictions of Economic Theory in Nonparametric Methods, in *Handbook of Econometrics*, vol. 4, R. F. Engle and D. F. McFadden, eds., Amsterdam: Elsevier, Ch. 42.

Melenberg, B. and A. H. O. van Soest (1996). Measuring the Costs of Children: Parametric and Semiparametric Estimators, *Statistica Neerlandica*, **50**, 171-192.

Meyer, B. D. (1990). Unemployment Insurance and Unemployment Spells, *Econometrica*, **58**, 757-782.

Müller, H.-G. (1984). Smooth Optimum Kernel Estimators of Densities, Regression Curves and Modes, *Annals of Statistics*, **12**, 766-774.

Murphy, S. A. (1994). Consistency in a Proportional Hazards Model Incorporating a Random Effect, *Annals of Statistics*, **22**, 712-731.

Murphy, S. A. (1995). Asymptotic Theory for the Frailty Model, *Annals of Statistics*, **23**, 182-198.

Murphy, S. A., A. J. Rossini, and A. W. Van der Vaart (1997). Maximum Likelihood Estimation in the Proportional Odds Model, *Journal of the American Statistical Association*, **92**, 968-976.

Newey, W. K. (1990). Efficient Instrumental Variables Estimation of Nonlinear Models, *Econometrica*, **58**, 809-837.

Newey, W. K. (1993). Efficient Estimation of Models with Conditional Moment Restrictions, in *Handbook of Statistics, Vol 11*, G. S. Maddala, C. R. Rao, and H. D. Vinod, eds., Amsterdam: North-Holland Publishing Company.

Newey, W. K. (1994). Kernel Estimation of Partial Means and a General Variance Estimator, *Econometric Theory*, **10**, 233-253.

Newey, W. K. and T. M. Stoker (1993). Efficiency of Weighted Average Derivative Estimators and Index Models, *Econometrica*, **61**, 1199-1223.

Nielsen, G. G., R. D. Gill, P. K. Andersen, and T. I. A. Sørensen (1992). A Counting Process Approach to Maximum Likelihood Estimation in Frailty Models, *Scandinavian Journal of Statistics*, **19**, 25-43, 1992.

Parner, E. (1997a). Consistency in the Correlated Gamma-Frailty Model, unpublished working paper, Department of Theoretical Statistics, University of Aarhus, Aarhus, Denmark.

Parner, E. (1997b). Asymptotic Normality in the Correlated Gamma-Frailty Model, unpublished working paper, Department of Theoretical Statistics, University of Aarhus, Aarhus, Denmark.

Petersen, J. H., P. K. Andersen, and R. D. Gill (1996). Variance Components Models for Survival Data, *Statistica Neerlandica*, **50**, 193-211.

Pettit, A. N. (1982). Inference for the Linear Model Using a Likelihood Based on Ranks, *Journal of the Royal Statistical Society*, **44**, 234-243.

Pollard, D. (1984). *Convergence of Stochastic Processes*, New York: Springer-Verlag.

Powell, J. L. (1994). Estimation of Semiparametric Models, in *Handbook of Econometrics*, vol. 4, R. F. Engle and D. F. McFadden, eds., Amsterdam: Elsevier, Ch. 41.

Powell, J. L., J. H. Stock, and T. M. Stoker (1989). Semiparametric Estimation of Index Coefficients, *Econometrica*, **51**, 1403-1430.

Prentice, R. and L. Gloeckler (1978). Regression Analysis of Grouped Survival] with Application to Breast Cancer data, *Biometrics*, **34**, 57-67.

Rao, C. R. (1973). *Linear Statistical Inference and Its Applications*, New York: John Wiley & Sons.

Ridder, G. (1990). The Non-Parametric Identification of Generalized Accelerated Failure-Time Models, *Review of Economic Studies*, **57**, 167-182.

Robinson, P. M. (1987). Asymptotically Efficient Estimation in the Presence of Heteroskedasticity of Unknown Form, *Econometrica*, **55**, 875-891.

Robinson, P.M. (1988). Root-N-Consistent Semiparametric Regression, *Econometrica*, **56**, 931-954.

Serfling, R. J. (1980). *Approximation Theorems of Mathematical Statistics*, New York: John Wiley & Sons.

Schucany, W. R. and J. P. Sommers (1977). Improvement of Kernel Type Density Estimators, *Journal of the American Statistical Association*, **72**, 420-423.

Silverman, B. W. (1978). Weak and Strong Uniform Consistency of the Kernel Estimate of a Density and Its Derivatives, *Annals of Statistics*, **6**, 177-184.

Silverman, B. W. (1986). *Density Estimation for Statistics and Data Analysis*, London: Chapman & Hall.

Stoker, T. M. (1986). Consistent Estimation of Scaled Coefficients, *Econometrica*, **54**, 1461-1481.

Stoker, T. M. (1991a). Equivalence of Direct, Indirect and Slope Estimators of Average Derivatives, in *Nonparametric and Semiparametric Methods in Economics and Statistics*, W. A. Barnett, J. Powell, and G. Tauchen, eds., New York: Cambridge University Press.

Stoker, T. M. (1991b). *Lectures on Semiparametric Econometrics*, Louvain-la-Neuve, Belgium: CORE Foundation.

Stone, C. J. (1980). Optimal Rates of Convergence for Nonparametric Estimators, *Annals of Statistics*, **8**, 1348-1360.

Stone, C. J. (1982). Optimal Global Rates of Convergence for Nonparametric Regression, *Annals of Statistics*, **10**, 1040-1053.

Stone, C. J. (1985). Additive Regression and Other Nonparametric Models, *Annals of Statistics*, **13**, 689-705.

Stuart, A. and J. K. Ord (1987). *Kendall's Advanced Theory of Statistics*, Vol. 1, New York: Oxford University Press.

Tsiatis, A. A. (1981). A Large Sample Study of Cox's Regression Model, *Annals of Statistics*, **9**, 93-108.

White, H. (1982). Maximum Likelihood Estimation of Misspecified Models, *Econometrica*, **50**, 1-25.

Index

Lecture Notes in Statistics

For information about Volumes 1 to 56
please contact Springer-Verlag

Vol. 95: Shelemyahu Zacks, Stochastic Visibility in Random Fields. x, 175 pages, 1994.

Vol. 96: Ibrahim Rahimov, Random Sums and Branching Stochastic Processes. viii, 195 pages, 1995.

Vol. 97: R. Szekli, Stochastic Ordering and Dependence in Applied Probability. viii, 194 pages, 1995.

Vol. 98: Philippe Barbe and Patrice Bertail, The Weighted Bootstrap. viii, 230 pages, 1995.

Vol. 99: C.C. Heyde (Editor), Branching Processes: Proceedings of the First World Congress. viii, 185 pages, 1995.

Vol. 100: Wlodzimierz Bryc, The Normal Distribution: Characterizations with Applications. viii, 139 pages, 1995.

Vol. 101: H.H. Andersen, M.Højbjerre, D. Sørensen, P.S.Eriksen, Linear and Graphical Models: for the Multivariate Complex Normal Distribution. x, 184 pages, 1995.

Vol. 102: A.M. Mathai, Serge B. Provost, Takesi Hayakawa, Bilinear Forms and Zonal Polynomials. x, 378 pages, 1995.

Vol. 103: Anestis Antoniadis and Georges Oppenheim (Editors), Wavelets and Statistics. vi, 411 pages, 1995.

Vol. 104: Gilg U.H. Seeber, Brian J. Francis, Reinhold Hatzinger, Gabriele Steckel-Berger (Editors), Statistical Modelling: 10th International Workshop, Innsbruck, July 10-14th 1995. x, 327 pages, 1995.

Vol. 105: Constantine Gatsonis, James S. Hodges, Robert E. Kass, Nozer D. Singpurwalla(Editors), Case Studies in Bayesian Statistics, Volume II. x, 354 pages, 1995.

Vol. 106: Harald Niederreiter, Peter Jau-Shyong Shiue (Editors), Monte Carlo and Quasi-Monte Carlo Methods in Scientific Computing. xiv, 372 pages, 1995.

Vol. 107: Masafumi Akahira, Kei Takeuchi, Non-Regular Statistical Estimation. vii, 183 pages, 1995.

Vol. 108: Wesley L. Schaible (Editor), Indirect Estimators in U.S. Federal Programs. viii, 195 pages, 1995.

Vol. 109: Helmut Rieder (Editor), Robust Statistics, Data Analysis, and Computer Intensive Methods. xiv, 427 pages, 1996.

Vol. 110: D. Bosq, Nonparametric Statistics for Stochastic Processes. xii, 169 pages, 1996.

Vol. 111: Leon Willenborg, Ton de Waal, Statistical Disclosure Control in Practice. xiv, 152 pages, 1996.

Vol. 112: Doug Fischer, Hans-J. Lenz (Editors), Learning from Data. xii, 450 pages, 1996.

Vol. 113: Rainer Schwabe, Optimum Designs for Multi-Factor Models. viii, 124 pages, 1996.

Vol. 114: C.C. Heyde, Yu. V. Prohorov, R. Pyke, and S. T. Rachev (Editors), Athens Conference on Applied Probability and Time Series Analysis Volume I: Applied Probability In Honor of J.M. Gani. viii, 424 pages, 1996.

Vol. 115: P.M. Robinson, M. Rosenblatt (Editors), Athens Conference on Applied Probability and Time Series Analysis Volume II: Time Series Analysis In Memory of E.J. Hannan. viii, 448 pages, 1996.

Vol. 116: Genshiro Kitagawa and Will Gersch, Smoothness Priors Analysis of Time Series. x, 261 pages, 1996.

Vol. 117: Paul Glasserman, Karl Sigman, David D. Yao (Editors), Stochastic Networks. xii, 298, 1996.

Vol. 118: Radford M. Neal, Bayesian Learning for Neural Networks. xv, 183, 1996.

Vol. 119: Masanao Aoki, Arthur M. Havenner, Applications of Computer Aided Time Series Modeling. ix, 329 pages, 1997.

Vol. 120: Maia Berkane, Latent Variable Modeling and Applications to Causality. vi, 288 pages, 1997.

Vol. 121: Constantine Gatsonis, James S. Hodges, Robert E. Kass, Robert McCulloch, Peter Rossi, Nozer D. Singpurwalla (Editors), Case Studies in Bayesian Statistics, Volume III. xvi, 487 pages, 1997.

Vol. 122: Timothy G. Gregoire, David R. Brillinger, Peter J. Diggle, Estelle Russek-Cohen, William G. Warren, Russell D. Wolfinger (Editors), Modeling Longitudinal and Spatially Correlated Data. x, 402 pages, 1997.

Vol. 123: D. Y. Lin and T. R. Fleming (Editors), Proceedings of the First Seattle Symposium in Biostatistics: Survival Analysis. xiii, 308 pages, 1997.

Vol. 124: Christine H. Müller, Robust Planning and Analysis of Experiments. x, 234 pages, 1997.

Vol. 125: Valerii V. Fedorov and Peter Hackl, Model-oriented Design of Experiments. viii, 117 pages, 1997.

Vol. 126: Geert Verbeke and Geert Molenberghs, Linear Mixed Models in Practice: A SAS-Oriented Approach. xiii, 306 pages, 1997.

Vol. 127: Harald Niederreiter, Peter Hellekalek, Gerhard Larcher, and Peter Zinterhof (Editors), Monte Carlo and Quasi-Monte Carlo Methods 1996, xii, 448 pp., 1997.

Vol. 128: L. Accardi and C.C. Heyde (Editors), Probability Towards 2000, x, 356 pp., 1998.

Vol. 129: Wolfgang Härdle, Gerard Kerkyacharian, Dominique Picard, and Alexander Tsybakov, Wavelets, Approximation, and Statistical Applications, xvi, 265 pp., 1998.

Vol. 130: Bo-Cheng Wei, Exponential Family Nonlinear Models, ix, 240 pp., 1998.

Vol. 131: Joel L. Horowitz, Semiparametric Methods in Econometrics, ix, 204 pp., 1998.